PROGRESS IN AGRICULTURAL PHYSICS AND ENGINEERING

Progress in Agricultural Physics and Engineering

Edited by

John Matthews

CBE, BSc, CEng, CPhys, FIAgrE,
FInstP, FErgS, FRAgrS, MemASAE

Former Director
Silsoe Research Institute,
Wrest Park, Silsoe,
Bedford MK45 4HS, UK

C·A·B International

C·A·B International   Tel: Wallingford (0491) 32111
Wallingford   Telex: 847964 (COMAGG G)
Oxon OX10 8DE   Telecom Gold/Dialcom: 84: CAU001
UK   Fax: (0491) 33508

British Library Cataloguing-in-Publication Data
A catalogue record for this book is available from the British Library

ISBN 0 85198 705 2

Typeset by Brian Smith Partnership, Bristol
Printed and bound in the UK by Redwood Press Ltd, Melksham

Contents

Contributors

W. BAADER, *Institut für Technologie (FAL), Bundesallee 50, D-3300 Braunschweig, Germany.*

BERNARD BAILEY, *Silsoe Research Institute, Wrest Park, Silsoe, Bedford MK45 4HS, UK.*

PIERRE BAYLOU, *École Nationale Superieure d'Électronique et de Radioélectricité de Bordeaux (ENSERB), 351 Cours de la Libération, 33405 Talence Cedex, France.*

D. R. FREITAG, *(Professor, Civil Engineering, retired, Tennessee Technical University), 312 Parragon Road, RR6, Cookeville, Tennessee 38501, USA.*

H. GÖHLICH, *Institut für Landtechnik, Zoppoter Strasse 35, D 1000, Berlin 33, Germany.*

TEMPLE GRANDIN, *Grandin Livestock Systems Inc., 2918 Silver Plume Drive, Suite C3, Fort Collins, Colorado 80526, and Assistant Professor, Department of Animal Sciences, Colorado State University, Fort Collins, Colorado 80523, USA.*

A. G. M. HUNTER, *Equipment Behaviour Section, Scottish Centre for Agricultural Engineering, Bush Estate, Penicuik, Midlothian EH26 0PH, UK.*

J. A. MARCHANT, *Silsoe Research Institute, Wrest Park, Silsoe, Bedford MK45 4HS, UK.*

R. ŘEZNÍČEK, *Katedra Fysiky, Vysoka Skola Zemedelska V Praze, Praha 6 – Suchdol 160 21, Czechoslovakia.*

Margarita Ruiz Altisent, *Departmento de Ingenieria Rural, Escuela Tecnica Superior de Ingenieros Agronomos, Universidad Politecnica, 28040 Madrid, Spain.*

Francis Sevila, *Division Génie des Equipements Agricoles et Alimentaires, CEMAGREF, 361 rue J.-F. Breton, B. P. 5095, 34033 Montpellier Cedex 1, France.*

R. D. Wismer, *Director, Research and Advanced Technology, Technical Center, Deere & Company, 3300 River Drive, Moline, Illinois 61265, USA.*

J. L. Woods, *Agricultural Engineering Section, Department of Agricultural and Environmental Science, University of Newcastle upon Tyne, Newcastle upon Tyne NE1 7RU, UK.*

Preface

Agriculture and other 'rural' industries, such as forestry, horticulture and aquaculture, as well as amenity area creation and maintenance, depend to a considerable extent on mechanized processes. Initially most agricultural engineering was concerned with crop production, but increasingly the husbandry of animals (including that of birds and fish) incorporates mechanical or physical, and even robotic, processes. Information capture and employment is very rapidly increasing to help optimize plant and animal system performance, and sensors, sophisticated process control systems, artificial intelligence and automatically guided machines all have a place in the present or the near future.

Alongside the revolution in biological sciences, based on advances in molecular biology, a similar, very rapid advance is being made in the physical and engineering technologies applicable to the biology based 'industries'. This is a result of the dramatic change in the power of computers and computing, which is reflected not only by the incorporation of computational components in machines and systems, but also by the use of computing to analyse, understand and develop their performance.

One result of this rapid advance is the increasing difficulty for engineers and scientists to maintain up-to-date awareness of even their own specialisms – and especially of topics outside their immediate field. Inter-disciplinary working and teaching are becoming increasingly vital, so that the need to understand what others are doing, or are able to do, is also increasing.

This publication brings together world experts to explain for the specialist, and the non-specialist, the recent advances in the understanding and application made in their individual subject areas, as well as the remaining problems and challenges. The information should be valuable (if not vital) to research engineers and scientists, to lecturers and teachers wishing to be up-to-date, to manufacturers of agricultural and associated equipment, to extension specialists and consultants, and even to those many technologically-biased farmers who have contributed so much to the advance

The scope of subjects is necessarily wide: mechanical, electrical, chemical, civil and information engineering are combined with several branches of applied physics. The common feature is the application to biological systems, where the inherent variabilities and complexity often make the engineering or physics particularly challenging. The experience, abilities and reputation of the chosen authors have ensured that these challenges are well addressed.

John Matthews

Chapter 1

An Analysis of Research Priorities in Agricultural Physics and Engineering

John Matthews

Introduction

Although a few decades ago agricultural engineering was considered a complete and easily explained topic – its title was self-explanatory – the introduction of newer technologies in the 1960s and the following two decades has led to much thought and debate about the scope of the activity. Agricultural engineering was predominantly based on mechanical engineering. Its graduates were taught this, plus some agriculture, underpinned by a very brief treatment of plant and animal science. The science and engineering of soil and water completed the planned study.

Increasingly, however, other branches of engineering have become important. Electrical engineering, incorporating micro-electronics, control engineering, robotics and information technology, has entered most application areas in agriculture. Chemical engineering must not only address problems of biological wastes and agrochemicals, but also be brought to bear on crops to add value and to make them appropriate for new uses. Civil engineering relates to the industry's use of soils – more soil is moved annually by agriculture than by what is labelled as the civil engineering industry. Most agricultural engineering involves a close interaction between equipment and biological materials – crops or animals – and for optimum process design this usually implies consideration of the 'engineering science' of the interaction.

Engineering science and applied physics virtually form a technological continuum, introducing physics inevitably into any attempt to define the scope and needs of today's agricultural mechanization. Physics has not, of course, only become relevant to agriculture in the last few years. Environmental physics, related both to the macroclimate and the microclimate, the physics of soil, physical properties of biological materials and heat and mass transfer,

1

have been recognized and successfully pursued for much longer. What has changed is the recognition of the need for a multidisciplinary approach. This is represented, for example, in engineering design being based always on a quantified understanding of the physical and mechanical properties of the processed biological material, together with the physics (or chemistry) of the process itself.

This integration of agricultural physics and engineering must not tempt its practitioners to draw anything but a boundary of convenience with biology. The fundamental objective is to establish, nurture and expand biological systems. Not only must the yield and economic efficiency of the system be optimized, but the needs of living organisms must be continually addressed. These will include the perceived needs of the community. These needs are related, for example, to farm animal welfare, environmental preservation, and the improvement of workforce health and safety. However, the biological sciences are, at the time of writing, in the early 1990s, advancing at a staggeringly fast rate. The potential and the exploitation of molecular biology, resulting in powerful opportunities in genetics and in biotechnology, mean that large resources are inevitably and quite properly devoted to the biology of agriculture. Engineers and physicists working to further the agricultural industry cannot hope to hear about or to understand all that is appreciated by the biologists; they certainly cannot properly judge the potential that the biologists can sense. The implication must be that agricultural physicists and engineers should understand enough about biological materials, organisms and systems to be able to interact with biologists. They must seek and ensure this interaction. If erring, it would be better to interact on an excessive rather than an inadequate level. It would be my judgement that the collaboration must be sought more by the physical scientist than will happen through the biologist, because the latter has less of a tradition of collaboration and also is inclined less to multidisciplinary system thoughts.

In completing this definition of 'agricultural physics and engineering' it is necessary to consider the scope of the application of the science and engineering on either an industrial or on a biological science basis. The above text should have clarified that a particular skill of the practitioners is their ability to apply physics and engineering to biologically based industries. Clearly, 'agriculture' should be taken, then, as a shorthand or generic title for the agricultural, horticultural, forestry and aquacultural industries. The last needs some comment, as in the 1950s and 1960s fish would have been considered outside our scope. Experience has shown, however, the very close similarity between the environmental provision, the feeding, handling and processing of farmed fish and that of agricultural animals and birds. The husbandry and processing of crops will imply their harvesting, handling and storage on the farm, and much of the processing of fresh crops − for example the trimming, wrapping and packaging of vegetables − will be an appropriate area of development for agricultural physics and engineering. Finally, we need to recognize that the environmental amenity sector, dealing with parks and gardens, roadside verges and large maintained countryside tracts, must come within our purview whenever a physical scientist's understanding of crop, soil or water processes must be involved. This, for the moment, defines the application area. New industrial sectors may well arise, based on biotechnology, and the physicist and engineer must take some responsibility for identifying and promoting the opportunities.

This book has been formulated by the research and development community. The state-of-the-art contributions will, it is believed, help that community to stay sufficiently informed on a broader basis about their unique discipline sector. All will not be read, but the chapters will stand as a reference, with analyses, views and predictions made by leading experts. These should provide insight for the judgement of an idea's potential as well as a starting point for new developments of thought or of hardware and software. Equally important, and the main reason for this paragraph, is the acceptance of the need for increased communication and understanding between the R&D sector and those in education, extension and the user industries. It is vital that the spectrum of these activities is seen substantially as a continuum. Educationalists must be aware not only of achievements and facts but also of possibilities and potentials. Advisers likewise must appreciate where understanding and technology has reached and where they are leading. There will be much in these chapters for the educated farmer and for leaders in the other industrial sectors.

In accepting that the sectors of industry in which we are interested are based primarily on biology, it is appropriate to question the comparative significance of the physics and engineering. The obvious prime criterion is the economic significance of the mechanization inputs to the industries. In more detailed analyses later it will be shown that, not only are these as high as 40% of total inputs, but engineering advances can very much influence other input costs.

Inevitably there will be differences in the comparative importance at any particular time of the various subjects − disciplines or applications − within the physics and engineering sector. Adequate analyses of the economics, including those of technological advance which can be hoped for, must be made. Socially based motivations will be important, particularly where advance is related to a perceived and well-publicized problem, such as nitrates in soils. This class of advance is frequently classified as 'Problem Pull'. Equal weight must be given, however, by those responsible for setting priorities or programmes to the availability of new ideas, new technological possibilities and new scientific and engineering skills to pursue the subject. With substantial international resources devoted to fundamental and strategic research as well as applied research and development, new possibilities for application are arising at a high and probably increasing rate. This source of advance is called 'Technology Push'. It is therefore important that an examination is made of where the Technology Push conditions most strongly apply, and it will generally then be more productive, and even economically advantageous, to seek new mechanisms, equipment or totally new systems in these areas.

In choosing subjects and authors for *Progress in Agricultural Physics and Engineering*, both Problem Pull and Technology Push criteria have been used. Some prediction has been attempted on which subjects, although not currently ready for rapid and fulsome application, are likely soon to yield profitable exploitation.

Reduction of production or sustaining costs

For product-based industries such as agriculture, measurements and calculations of costs per unit output can be made relatively easily. For the environmental area − including parks and gardens − this is clearly less practicable. Input costs can be measured, but there is no output to be quantified other than a judgement of desired quality of the environment made subjectively.

For production enterprises, technological advance must be associated with one of the following benefits if it is to be justified.

1.   *A reduction in the costs associated with equipment, buildings or labour.* This could include a direct reduction in the cost of constructing a machine resulting from a redesign, use of a new material, or a new mechanism. It could come from reduced maintenance costs from a better design, from increased machine output per unit of time, or a lowering of labour costs. The last will occur if the machine can carry out the task more rapidly or if the crew or associated workers are decreased in number. An automatic feeder would obviously reduce labour.

2.   *A saving of materials which constitute the other principal input costs of the industry.* In agriculture these will be seed, fertilizers, protective chemicals, animal feedstuffs, veterinary pharmaceuticals and fuels. Good examples would be spraying equipment which incorporates better targeting of chemicals, allowing lower quantities to be used, or an improved animal feeding system which, through the use of growth/nutrition models, optimizes and accurately controls the quantity fed.

3.   *An improvement in the efficiencies of growth of plants or animals.* Such might come about for plants through the improved control of the seedbed, the water supply or the greenhouse environment. For animals it has been shown that improvement to the quality of air in a piggery, by filtering out the dust particles and any attached pathogens, can lead to faster growth rate, apparently by improving the feed conversion rate in the healthier animal.

4. *Improving or preserving the quality of the product.* Quality can be influenced by processes during growth, from preparation of an optimum seedbed to crop protection. At and after harvest the crop inevitably undergoes much handling and, often, storage. There is great potential for damage and hence quality reduction, both in handling and transport and in storage. A good understanding of the mechanics, of the physics, and of the processes can lead to a better maintenance of quality.

5. *Increased reliablity of supply and extension of the season.* This may be achieved by, for example, access to the harvesting of a crop which is less dependent on good weather or soil conditions. The replacement of tractors by gantry-based equipment to harvest cauliflowers causing less soil rutting and crop damage in wet weather, is a good case.

6. *The improvement of a technique or system to make it more socially acceptable.* Although at first sight this category of advance may appear totally based on moral grounds and devoid of economic grounds, financial considerations are not totally absent. This arises because, if the public become convinced that a working practice or its result is unacceptable, they are likely through government action or local pressure to seek its suspension or revision with a consequent cost penalty.

Relatively little has been published that analyses production costs on a worldwide basis. Many national data exist, however, and recently the European Community Club of Advanced Engineering for Agriculture (ECCAEA) has published an analysis of relationships between mechanization costs, total input costs and gross agricultural output values in five of the major countries of the Community. In relation to input costs the mechanization component is quoted as varying from 36% in the UK to 47% in Italy (see Table 1.1). Unpublished analyses in the UK have been carried out on individual agricultural enterprise types. For winter wheat,

Table 1.1 Relationships between mechanization costs, total input costs and gross agricultural output values in some EC countries.

Countries	Proportion of mechanization costs to gross agricultural output (%)	Proportion of mechanization costs to total input costs (%)
France	21	38
Germany[a]	28	44
Italy	23	47
Netherlands	29	45
United Kingdom	24	36

[a]Pre-unification (German Federal Republic)

mechanization and associated labour costs (buildings excluded) amount to 31% of total costs; for milk production, equipment, buildings and labour account for 50% of costs; in tomato production, the cost of buildings, fuel and labour is as much as 78% of total costs.

More important is the potential for a significant reduction in these costs. It is relatively easy to calculate the theoretical economic advantages of one system component change in isolation. Such 'theory', however, often ignores many factors, including the inability to make small changes in labour employment, where any real saving must normally be one complete person. Interaction between system components and inefficiencies during the learning of new techniques or systems are also ignored. It has been pointed out that in some cases when claims are made for individual system components and these are then summed, the system as a whole is alleged to be able to run at a negative cost! Nevertheless, such calculations are valid for comparisons of priorities and, where operational research modelling is employed with an adequate rigour, absolute values for cost-saving potential can be defended.

Relatively simple estimates and calculations of the potential for cost reduction through mechanization advance have been undertaken by a working group from six European countries, led by Pellizzi, and sponsored by the EC Agriculture and Energy Committee. Although originally studying the potential for energy input reduction, this group recognized the need to judge and compare possibilities on an economic basis. The group studied several areas including tractor use, tillage, harvesting, forage and biological waste products. In Table 1.2 the assessed cost-saving potential in the growing of cereal grains, published from these data, is presented. A total potential of almost 25% is estimated which, although perhaps suffering from some of the reservations listed above, is nevertheless large enough to be very important in, for example, furthering the use of cereals in non-food applications.

Table 1.2 Estimated potential input cost reductions for cereals achievable by engineering technology advance.

Operation or materials	Potential saving in cost of operation and materials (%)
Cultivations	50
Drilling and fertilizing	7.5
Spraying	22
Harvesting and handling	50
Drying	5
Total operating costs	24

A similar study has been undertaken in parallel with and independently of the EC Group, but with the input from a number of individual operational research (OR) and other determinations made at the AFRC Institute of Engineering Research, UK. This led (Table 1.3) to a prediction of 10% cost saving in the next decade and a 20% total potential over the longer 20–25 year period.

Table 1.3 Estimated mechanization-related costs of wheat production in the UK, together with judged potential for reduction through engineering advance.

Item	Cost (£/ton)	Potential savings, 1990s	Potential savings year 2000 onwards
Plough	5 ⎤		
Disc harrow	4 ⎬	2	6
Seedbed	1 ⎦		
Drill	1	–	–
Roll	1	–	–
Fertilize	2	–	–
Spray (3)	3	1	1
Combine	9	2	4
Cart grain	2	–	–
Dry	8	1	2
Bale straw	1	–	–
Cart straw	2	–	–
Subtotal	39	6	13
Seed	6	–	1
Fertilizer	13	1	3
Sprays	13	3	4
Subtotal	32	4	8
Total cost	71	10	21

On the broader agricultural basis a number of individual mechanization R&D projects have been subject to economic analysis and these were reported in Matthews, 1988. The results are summarized in Table 1.4 and indicate an ample justification for improving the technology. They also fall into the different classes of economic justification listed above.

Finally in this section, it is important to look at the technological resource from which the advances can come. In this review I will draw on the publication of ECCAEA and its document 'Agriculture and Engineering – New Technological Opportunities' to which I played a contributing part.

Table 1.4 Estimated benefits arising from a variety of current engineering developments.

Development	Estimated benefit
Robotic milking	Annual saving of £35–£60 ($70–$110) per cow
High-technology sprayer	On a 400 ha arable farm, more than £10,000 ($19,000) per annum saved
Filtration of air in piggery building	Net saving per pig of £1.06 ($2)
Improved environment in mushroom-growing house	A net yield gain of 3%, coupled with reduced inputs
Incorporation of a specialized gantry in cauliflower growing	Possibility of 20% increase in crop yield, this more than justifying extra capital cost
Suspension incorporated in tractors	Net saving of £600 ($1,100) per annum

Eight technology areas are listed and in each case the topic is illustrated by a few examples.

Sensors

New devices and improved, miniaturized and cheaper signal-processing components are rapidly widening the scope for measurement. Biosensors, electrochemical sensors and devices employing many physical principles, often allowing non-contact measurement, feature strongly. Combinations of sensors to indicate more complex phenomena or to provide, for example, automatic temperature correction, are increasingly employed.

Examples of sensor applications are:

- crop parameters affecting production and quality;
- soil and seedbed characteristics;
- controlling and monitoring field machines with increased precision;
- livestock parameters related to welfare and quality;
- process monitoring and control in the biotechnology and food industries.

Vision systems and image analysis

Organism location and recognition, analysis and characterization of complex scenes and, ultimately, the replacement of human observation and monitoring of livestock by computer-based machine vision are all possible. Inherent

biological diversity and variability make the application of vision systems to agriculture and biologically based industries particularly challenging.
Examples of vision system applications are:

- actuator guidance in robotic systems;
- characterization of biological materials, for example for quality;
- soil/seedbed characterization;
- detection of weeds, pests and diseases;
- livestock monitoring for health and growth.

Robotics

Robots will be most attractive where they can replace arduous or repetitive manual tasks. The seasonal nature of many agricultural tasks may limit the economic viability unless multi-purpose robotic systems can be developed.
Examples of early application of robots are likely to be:

- milking of diary cattle;
- produce harvesting, particularly in horticulture and orchards;
- dissection of micro-plants;
- shearing of sheep.

Mathematical modelling of processes and systems

The advance related to modelling is a valid 'new technology area', because the availability of more powerful computers has led to a massive increase in our ability to model with validity and to cover more complex processes and systems – even taking into account the biological variability. We can contemplate valid enterprise management models in addition.
Summarizing through examples, these might be:

- process control models for on-line control;
- modelling of biological systems, such as heat and mass transfer in crops, to allow optimum design of driers;
- operational research and simulation models for the solution of management problems.

Information technology

Information systems and artificial intelligence should show particular benefit to the biologically based industries because human judgement in today's methods of management, although often extremely skilled and based on generations of passed-on experience, has nevertheless been shown to be very variable between practitioners, with many clearly making non-optimum judgements.

Examples of what can be expected of IT are:

- access to data banks, for example on farm machinery or market prices;
- expert systems to aid decision-making;
- electronic communication systems to couple sensors, artificial intelligence and actuators;
- user-friendly man/machine interfaces, particularly in the use of computer screens.

Process analysis and control

Process control will become of increased importance as more on-farm processes are developed for new crop uses and for 'added value' reasons. Process control is becoming more valuable as increased sophistication develops; for example, adaptive control, self-learning control or linked control systems.

Application examples are:

- the processing of crops for industrial use;
- new biotechnological processes;
- the control of processes such as handling, mixing or drying;
- waste processing.

New engineering science methodologies

New analytical techniques, often based on numerical methods or simulation, have allowed deterministic and stochastic processes to be better represented and understood. Typical techniques falling into this class are computational fluid dynamics, 'random walk' theories for particle transport, or finite element or difference methods applied to three-dimensional processes.

Application examples are:

- optimization of crop storage environments;
- removal of dust from housed livestock environments;
- reduction of drifting spray chemicals.

Soil mechanics and soil physics

Although this may be perceived as more of an application area than a new technology area, this topic is included partly because of the recent and relatively severe problems relating to the soil (nitrate and agrochemical pollution as well as compaction and erosion), and partly because of the advanced understanding allowed by computer modelling.

Application areas of highest priority for soil mechanics and physics are:

- development of more efficient soil processing (tillage) techniques;
- fluid transport processes in relation to water and chemicals;
- development of non-soil growing media;
- maintenance of the structure of agricultural soils and those in the natural environment.

Agriculture and the environment

This section will show that engineering and the physical sciences have an essential role in the protection of the rural environment, both in industrialized and less developed countries. In introducing this role, however, it is appropriate to recognize that, to many, the contribution that mechanization has made to agriculture has had the effect of greatly harming its environmental influence. Large open tracts with hedges removed, industrial-scale buildings, mud on rural roads, unsightly ruts in field gateways and lanes, noisy and dust-producing machines and slow convoys of cars behind agricultural vehicles are all common complaints. Most of the complaints can only be defended on grounds of plentiful and affordable food. The engineer's policy towards improving the rural environment must therefore encompass an improvement of many existing systems. I must also, however, include the derivation and application of engineering, physical or chemical solutions to many other problems which do not arise from mechanization but usually from farm intensification. I refer here to livestock wastes, nitrates in soils and pollution from crop protection chemicals.

Agrochemicals fall into two main classes, nutrients and pesticides (this word will be used to include herbicides, fungicides and insecticides). Nutrient chemicals or fertilizers have enabled farmers to feed large populations efficiently, but have sometimes been applied indiscriminately and in all cases there is still potential for more efficient use. Three factors can be readily identified: timing of application (including the decision whether to apply or not); determination of optimum quantity; and delivery of that quantity, including its spatial distribution. Advances in information technology (IT), incorporating crop-growth modelling and expert systems applied to husbandry, will provide useful and increasingly accurate support to the farmer on when and how much to treat a crop. Such decisions may be based on predictive models, but are more likely to provide the necessary accuracy when sensors can be employed to detect and measure nutrient deficiency. Many of these sensors still have to be developed, however. Having determined the needs, there is still much scope for improved placement encompassing either accurate evenness of spread over the whole surface or, for row crops, localized placement in relation to the plant and its roots. Meteorological interference, particularly from winds, must be made less influential.

The same technologies will be employed for systems applying pesticides. IT decision support has a predictive and judgemental role, but the predictive element is unlikely to be accurate without 'on-line' sensing. If the presence of weeds, disease or insects can be detected and quantified, then IT systems can help to judge whether spraying at that time or at a more optimum time will be economically effective or even environmentally adequately safe. Sensors to provide the necessary detection are being developed and, in parallel, biologists are studying the effectiveness of varying application concentrations. The design of applicators for pesticides is more important than that for nutrients since badly targeted or drifting pesticides are much more than an economic loss: they are a potentially very harmful influence on the local environment. The physics of particle transport is receiving a great deal of attention and undoubtedly useful process modelling will soon be able to be applied – together with relatively new principles such as air assistance and electrostatics – to spraying machinery. This is not yet possible, however, although by various means and in particular circumstances, substantial increases in deposition at the target, and hence potential reduced quantities of application, have been reported. An example is a 50% improvement in targeting in apple trees by the employment of electrostatic charging.

Agricultural by-products, mainly livestock wastes and crop residues, have with intensification presented quite severe environmental problems. Systems and equipment for the handling, treatment or disposal of these waste products will principally involve the chemical engineer, although mechanical and control engineers have a considerable role. Livestock waste treatment to remove malodours and to modify the consistency for ease of handling involves multidisciplinary work to encourage bacterial processes, most commonly by the supply of oxygen through aeration. Anaerobic treatments have been devised and employed on a restricted scale, mainly to preserve the methane produced as a fuel source, but many have found that for agriculture the economics are not viable. This is particularly true since waste production and fuel demand are both frequently seasonally variable. Consistency variation can be in two directions; to dilute and hence reduce the viscosity so that material may be pumped, or to separate solids which contain much of the nutrient value, so that they may be handled and transported as a solid.

Odour removal or reduction and the reduction of potentially harmful gases, notably ammonia, are not limited to the waste effluent. The escaping or expelled air of an intensive housing for farm animals can be unacceptable. Measures to reduce these components within the building, to filter or react them on egress, or to control the efflux atmosphere away from local populations, are all relevant. The contribution of harmful gaseous components from agriculture to global problems of acid rain or global atmospheric warming, is accepted in some countries as substantial. Ammonia and odours arise in the most concentrated form from the disposal through spreading of manure slurries. Research and development on liquid trajectories during spreading, on means of rapid absorption into the soil, on reducing re-evaporation through the use of crop cover, and on tillage treatments to cover the effluent, are all relevant. In all treatment and system considerations, the nutrient chemical value of the material is an important factor which is likely to increase in importance as recycling of chemicals gains priority under 'green' pressures.

Crop wastes will also gain priority as by-products to be employed wherever practicable. Whole-crop utilization systems with feed, industrial and fuel components, are being developed, particularly for the time early in the next century when the economics of petroleum-based products are less attractive. Many 'waste' components, particularly plant roots or partially decomposed outer or lower leaves, for example, will nevertheless probably serve best as humus for soil structural maintenance. In some countries the most problematic waste component is straw from cereals; burning, a commonly utilized system for some decades, has been banned. The incorporation of straw into the soil is a practicable alternative, but should perhaps be seen as a transient one in view of the substantially increased energy expenditure in tillage which is deep enough to allow this incorporation. Straw is a good source of fibre and cellulose, but their extraction is currently uncompetitive. In considering overall environmental and bio-industry options, the needs for adequate and appropriate markets for forestry products must be balanced against those of agriculture.

Other pollutants include dust and noise. Dust from crops and soil processes can lead to seasonal complaints. Harvesting of deteriorated cereal crop can result in clouds of dust drifting away from the combine harvester to affect local residents. Seedbed preparation on dry soils can lead to similar phenomena. These would generally be seen as less serious problems than those of odours, but they may need to be ameliorated in the next few years through study of the environmental physics and of improved machine design. Noise from tractors is restricted by regulations designed to limit noise levels arising from road traffic. These regulations probably provide adequate environmental noise protection from the tractor. Other sources, not subject to regulation, but which have led to serious complaint, include tree and bush spraying equipment with high frequency sound components from fans, and crop-drying equipment, again predominantly from the fan blades. The former source is intermittent in that trees and bushes are sprayed and then left for a few days or weeks. They tend to occur, however, when the weather is warm and rural residents are outside or have windows open. Crop-drying plant can, in the author's experience, lead to cases of annoyance at quite large distances – perhaps up to 1 km from the source. This arises from a combination of operational hours, often including the night and at least the evening and weekends, together with focusing of the sound by topographical features.

Agricultural soils, in both arable and grassland states, as well as those constituting lanes and tracks, are suffering from compaction and from erosion. Both result from structural damage which is partly a result of intensive cropping and partly caused by the passage of increasingly heavy traffic by tractors and machines. These machines and agriculture itself cannot of course be blamed for all erosion problems. It is also a natural phenomenon caused by weather and soil type. Nevertheless the erosion problem is a universal one represented both by blowing dust storms in areas such as North America and by soil carried by almost all rivers in Africa as evidence of spreading agriculture. The engineer's and soil scientist's roles are first, to understand better these structural change processes, to quantify them in terms of mathematical models that will describe traffic or tillage processes and their effects, and then to design new, lighter vehicles or ground drive systems which will cause less pressure or shearing stresses. New tillage devices are required that will produce fewer 'pans' or fewer aggregates that are too small. The presence of slopes, variability in field soils, efficient irrigation or drainage systems are all relevant.

Finally, the visual environment needs to be mentioned. The engineer can design better equipment for the maintenance and visual improvement of hedges and ditches, including both sprayers and fertilizer distributors that will not affect the more natural environment in the hedgerow. Reductions in rutting, particularly in gateways, will improve the view. The other major visual factor is the farmstead with its buildings, silos and parked machinery. The design of buildings and silos, as well as their location, arrangement and colouring, can all improve the view.

The safety and welfare of farm workers

The application of the subject of ergonomics or human engineering to agriculture and associated industries has been substantial since the end of World War II. The topic may be conveniently divided into environmental hazards and influences; workplace design; work analysis and task design; and, not least, accidents.

Environmental components of considerable importance to workers are noise, vibration, dust, fumes, climate and odours. Tractor operator noise is subject to limiting regulations in most countries and any workplace noise to daily exposure limits in some. Noises specific to our application sector and in need of some engineering solution include that from mills and feed-processing plant, from chain and circular saw blades and from certain conveyor systems; there is evidence that squeals from pigs may exceed safe limits. Vibration sources of the most importance are tractors and other unsuspended vehicles, where whole-body vibration and jolting commonly exceeds published human acceptability limits; and chain saws together with some powered hand tools, where hand–arm vibration levels again exceed published limits and can lead to damage to the vascular nervous system. Dust is the environmental nuisance most quoted by farm workers. As well as its discomfort effect, it can include pathogens from moulded plant material – mainly from hay – which can be responsible for the disease 'Farmers' lung'. Fumes and noxious gases can be fatal hazards in pits holding slurry and in silos with stored crops. They can appear at a dangerous level when combustion-engined machines are operated in farm buildings including greenhouses. The problem of odours for the farmer or farm worker is that some of them can be so concentrated that they penetrate not only the clothes, but even the skin, to remain evident long after work is finished for the day and to become a social liability.

Workplace design is often a severe challenge to ergonomists because:

1. it is necessary to observe both the route ahead and the work being done behind the machine;
2. a large number of separate controls is needed for the many functional components on most machines;
3. access to the workplace will often be difficult both because of muddy, uneven ground and because of highly mounted workplaces;
4. mud, rain and dust may often be present on glass windows;
5. the control tasks and any instrument or work monitoring must be undertaken whilst the vehicle is vibrating quite severely.

Work analysis and design have increasingly been targeted to more mental and intellectual tasks and away from essentially physical tasks requiring a large energy input. Nevertheless, the physical energy requirement of many tasks in less developed countries, such as those that involve handling implements hauled by draught animals, can still be near to the unacceptable. Consideration of the total system on a physiological energy input basis – both person and working animal – is called for. In industrialized countries the emphasis for scientific study and technological advance has shifted from tasks with a strong metal workload component, such as parlour milking or complex control tasks, to consideration of the 'man/machine' interface through computer and VDU interaction.

Accident hazards are inevitably high in agriculture and associated pursuits for several reasons:

1. the terrain and environmental conditions are poor, reducing personal stability;
2. many workers change tasks with the seasons, losing familiarity with and practice at what they are doing;
3. many workers operate alone, worsening the possible consequences of an accident in that help may not be available;
4. children and young people, as well as sometimes very old people, are involved as family members who nevertheless work.

The various determinants of accident risks are amenable to study and understanding by ergonomists, applied psychologists and human physiologists. The solutions to ameliorating risk or effect will lie with the designers of equipment or systems as much as with educators, trainers or supervisors. Workplace design, including access, seating or standing arrangements; control and monitor design; needs for any exertion, be it in control operation or object lifting and carrying; feedback of information; vision of a task or route; clear allocations of responsibilities between team members; and well-designed instruction manuals, are good examples of the more important factors. There is, in the author's view, an inadequate quantification of most of these factors as far as the industries we support are concerned.

Farm animal welfare

Socially motivated demands of the population as a whole for livestock to be better regarded and treated vary in intensity from country to country, but appear to be increasing everywhere. The closeness and apparent boredom of confinement, together with perceived stresses in transport and handling, are the principal complaints. Those concerned professionally and scientifically with livestock husbandry or animal science have identified many more needs. There is, however, a substantial scarcity of information on needs and preferences of animals, particularly in the quantification or relativities of the various factors. As an example, although a great deal of research has been carried out on human tolerance to vibration and various standards of acceptability have been agreed there has been no quantification of the stress caused or the acceptability of vibration applied to animals. This is despite the fact that during transport, vibration levels are known to be high and fatalities have been experienced with animals, birds and fish. Although this section is concerned with welfare, the interrelationship between welfare and productivity of animals is worthy of mention. It is not an unreasonable assumption that an animal which is contented, or at least relatively unstressed, will thrive better and produce either more meat or more milk than one less well treated. It could be argued that productivity is a measure that can be used to quantify welfare, but this may not be totally reliable.

The concept of welfare itself covers both physical and mental well-being and is also considered to include behavioural aspects, since it is accepted that animals should not be deprived of the opportunity to express natural patterns of behaviour. Health, or the absence of disease, is obviously one of the most basic indicators of welfare. The quantification of what level of stress may be regarded as acceptable is difficult, particularly as it may be argued that animals in a more natural, wild habitat will suffer stress from predators, food shortages and the like.

It is relevant to consider the contribution which engineers and physicists can make to welfare evaluation through devising monitoring equipment, including sensors, for the measurement of physiological or behavioural patterns. For example, mobile instrumentation packages to measure heart rates, body temperatures, and activity such as stepping or movement frequencies, are all proving valuable. Video recording and image analysis are increasing our ability to undertake detailed studies. Multidisciplinary research is essential and, as well as analytical conclusions, it can lead to expert systems and the use of artificial intelligence to improve husbandry.

The housed environment of farm animals (or the water environment of fish) has two main characteristics of importance: the quality of its cleanliness in terms of dust or dirt particles or of pathogens in suspension; and the climate characteristic, largely related to temperature. Air quality has been shown to affect growth and health and there is a parallel with sediment-free water. Optimum temperature for the apparent comfort of pigs as evidenced by their behaviour, and for their growth rate maximization, is known to be related to the animal's size and food intake, and to the interactive climatic factors, such as air movement. The physical environment incorporates in addition the nature and cleanliness of the wall surfaces and of the floor. The correct choice of constructional materials and of cleaning and maintenance systems must be part of the engineer's responsibility. Adequate illumination levels and even the absence of glare and shadows will influence welfare. As with the other environmental factors, good lighting will benefit the stock worker, improving both comfort and safety.

Good design of buildings or internal enclosures must encompass their size and shape to assist the animal's movement, feeding or interaction with fellow animals. Floors must minimize foot skidding or damage, and doorways, races or internal obstacles must not constitute bruising or abrasion hazards. Livestock 'ergonomics' is largely an unexploited topic.

Livestock equipment where design requirements are most important include milking machines, shearing units and, more recently, mechanized means of collection and transport for poultry. The interaction between milking equipment and mastitis is well known and analysed. Robotic attachment of the milking cluster should allow more frequent, or even on-demand, milking with signs of improved health and welfare. Robotic shearing seems unlikely to show welfare benefits, unless greater precision can avoid any small flesh cuts. Mechanized collection of broiler poultry has been shown, by heart rate measurement, to be less stressful than current manual methods which include carrying by the legs and hence bird inversion.

Transport of livestock has been mentioned earlier as a source of vibration stress. Losses occur, particularly with young creatures. The transport by land, sea and air of young fish − salmon smolts, for example − is an important example of the technical challenge. Movement and processes, including stunning and slaughter, at the abattoir or processing factory encompass many potential stresses. Considerable attention has been given to the subject and it is reviewed in this book. The review illustrates well the principles of employing natural behaviour in designing for movement and handling.

The quality and safety of produce

Quality may be defined as the possession of those characteristics which the customer desires. Engineering and the physical sciences are important in ensuring high quality in crop and animal products, throughout both the growing period and the subsequent processing. Quality in both crop-derived and animal-derived food refers to the sensory factors such as flavour, aroma and texture, as well as to cleanliness and the absence of blemishes. It must also imply safety and fitness for human consumption and requires an absence or sufficiently low content of potentially harmful micro-organisms, toxins and foreign bodies. Having achieved good quality and safety, subsequent handling and storage must as far as possible maintain that quality. There is usually a requirement for inspection, sorting or grading to separate produce into narrower quality categories.

Crop establishment affects not only overall yield, but also quality and uniformity of a crop. New physical methods are being devised to assess the germinative capacity of seeds, the quality and viability of plants for transplanting should be assessable using image analysis, and robotic systems can increasingly select or reject items. Models and expert systems can help to manage crops optimally. Chemicals used in crop protection are increasingly being seen as a food safety concern, perhaps wrongly, but their replacement by mechanical methods or even physical techniques, if these can be identified, would be highly popular.

Harvesting systems could be developed still further to reduce mechanical damage, whilst selective harvesting can give a quality boost, perhaps incorporating non-destructive maturity monitoring to optimize harvest date for individual items. Similar non-contact monitoring is needed for sorting apparatus, which should further evolve to increase speeds and yet reduce mechanical impacts or pressure stresses. Detection and removal of extraneous objects (stones, pests, unwanted stalks, leaves or stems, and perhaps even toxic weeds) would have wide application with a whole variety of crops.

Environmental applications range from storage facilities with closely controlled temperature, humidity and gaseous atmosphere, to rapid in-field or in-store cooling using vacuum. Future developments are likely to include 'intelligent' stores, in which automatic sensing systems determine changes in the state of the produce and, with a knowledge of marketing dates, adjust the atmosphere in such a way that optimum maturity is reached at the correct time. For cereals, intelligent drying plant should be able to minimize energy, maximize bread-making quality or germinative quality, and achieve optimum cooling for storage, all at the same time but with appropriate compromises.

In livestock production, quality partly depends on ensuring avoidance of undue stress during the period between farm and slaughter, since this is known to impair meat quality. Bruising or even bone damage can arise from handling or transport, again resulting in the downgrading of quality. Quality in animal products such as milk or eggs requires engineering for hygiene, and to improve monitoring techniques that prevent defective products, such as those from sick animals or those containing drugs, from entering the food chain.

Ensuring the safety of produce for human consumption still demands a range of new or improved inspection or test techniques to identify anti-nutritional components, pathogens or biochemical products associated with spoilage. New preservation techniques may be developed in association with food scientists. The modelling of biological deterioration processes may also benefit from such collaboration between the disciplines. More strategic research may show further useful correlations between measurable physical parameters and the quality factors, to lead to new, rapid sensing techniques for the biochemical or microbiological changes.

Conclusions

Inevitably this review is a somewhat personal one, with the topics chosen and weighting given being those of the author. The review has drawn on consensus from many national and international discussions, however, particularly those related to the Agricultural and Food Research Council in the UK, to the European Community Club of Advanced Engineering for Agriculture, the Club of Bologna, and bilateral discussions within several European and other countries.

What is indisputable is that engineering and the physical sciences have a major and very broad role in developing agriculture and associated biology-based industries to the environmentally and socially acceptable level that the population desires. Economic advances are also practicable and substantial. The expansion of markets for biological products to encompass industry feedstocks and fuels, as well as food, underlines the importance of reducing production costs.

Reference

Matthews, J. (1988) Chips with everything. *Proceedings of the Oxford Farming Conference*, UK.

Chapter 2

The Mechanics of Soil–Machine Interaction

R.D. Wismer and D.R. Freitag

Introduction

The term 'mechanics' carries several implications. First, it requires that the relations involved be described mathematically. Secondly, the configuration of the system and its motion must be expressed in concise geometrical terms. Thirdly, it is necessary for there to be equations that describe the behaviour of the soil material when subjected to the stresses imposed by the machine. Fourthly, the properties of the soil that are the parameters of the behaviour equations must be identified and measured. Finally, there must be an expression that predicts the result of the interaction of the machine with the soil. Both the physical and the philosophical aspects of the development of a soil–machine mechanics are discussed in some detail by Gill and Vanden Berg (1967).

Most soil-working equipment has been evolved through a trial and adjustment process over many years. Developments have been made by individual inventors with a vision of a better way of doing a task. Often their visualization was not realized and new trials were made. The trial-and-error method has been used because there was no way of stating exactly what forces and conditions must be employed to achieve the desired final state of the soil. In fact, there was no widely accepted way of stating what that end product was. However, in the last 50 to 100 years the search for better, more efficient machines has been aided by a steadily improving understanding of the mechanics of the interactions between the soil and the instrument or tool applied to it.

Although some progress has been made, there is much yet to be accomplished. A major problem is the extremely complex and variable nature of soil. Soil is not an entity like water. Water, wherever it is found, has essentially the same properties, and variations can be predicted by knowledge of the pressure and temperature. Soil exists in a myriad of varieties that can be described only by evaluating each one. Thus, even to measure the characteristics of a soil involves the concepts of probability and statistics. Furthermore, the measures used to portray the characteristics must be relatively simple and amenable to use under field conditions.

The tasks that must be done then are as follows.

1.   To reduce the infinite variety of soils to a finite number of groups to permit discussion and information exchange.

2.   To describe in precise quantitative terms the properties and characteristics of any soil relevant to its response to a machine-applied force system.

3.   To relate these soil properties and characteristics to the geometry of the machine to furnish a quantitative evaluation of:

   (a) the change that the soil will experience as to its displacement and physical characteristics, and

   (b) the forces and energy required.

4.   To establish methods of acquiring the necessary soil properties and characteristics quickly and effectively for large areas of the size associated with the activity under study.

Soil as a material

Particulate nature

Soil consists of pieces of the rocks and minerals of which the earth is composed. Some of the pieces are simply broken fragments of the base rocks, while others are clay minerals, the products of chemical weathering of the rocks. From a mineralogical standpoint, soils represent all of the diversity of the world's geology. Fortunately, except for the clay minerals, a piece of one kind of rock in a volume of particles responds to moderate levels of stress very much like a geometrically identical piece of another kind of rock.

Clay minerals are usually extremely small; small enough to have the properties of a colloid. The mechanical behaviour of a volume of clay particles is influenced chiefly by the inherent forces that arise at the molecular and atomic level. These forces strongly attract water, free ions and non-crystallized chemicals to the surface of the particles. In the presence of adequate water, a volume of soil containing an amount of clay has the properties of a plastic solid. If the water is lost, the particles are bonded by the adsorbed chemicals and the inherent forces, and the soil volume takes on the properties of a rigid solid. Upon rewetting the former condition is largely, but not completely, restored. This is the familiar seasonal transition of a soil from springtime mud to summer's brick-like hardness.

Multiphase system

A soil is composed of an assemblage of discrete particles of various shapes, sizes and orientations with the void space between the particles containing water or air, or both. The manner in which the assemblage is produced and its stress history determine the density of packing of the particles and the relative proportions of the three possible components. Natural soil-forming processes produce soil densities somewhere between the maximum and minimum possible states of the soil. Usually, the density is much nearer the minimum. Soils that are below the water table are considered to have all the voids filled with water. Other soils can have any ratio of air and water in the voids. In arid regions, with consistently low relative humidity, some water remains adsorbed on the particle surfaces, but for practical purposes, the voids are air-filled.

The amount of water in soil is the variable that has the greatest influence on its behaviour. This is particularly true for fine-grained soils. The forces that operate at the surface of the clay minerals strongly attract water molecules, ions and other clay minerals. These conditions are responsible for the characteristic properties of any soil with a significant proportion of clay. Such soils exhibit true cohesion, shrinkage, swell, plasticity, and high dry strength. When the voids are water-filled, the behaviour of these soils follows a fairly predictable pattern. When the voids are less than saturated, the effects of the third phase, the air, add a complexity that has not yet been adequately understood. Coarse-grained soils are not greatly affected by the presence of water. The exception is when the soil is unsaturated and the capillary forces of water at the points of contact of the individual particles create an apparent cohesion.

Effective stress

When a volume of particulate material experiences strain due to applied stress, the individual particles tend to move relative to one another. If the initial particle density is low, the strains allow the particles to move into void spaces, decreasing the volume and thus increasing the density. If the initial particle density is high, some of the particles will be forced out of position increasing the volume and decreasing the overall density. There is a 'critical density' at which the net effects of the individual particle movements result in no change in density.

When a saturated soil is stressed, the tendency for increase in density due to the shearing strains can be accomplished only by the expulsion of water from the void space. In coarse-grained soil, this can take place relatively rapidly, but for fine-grained soil the process is very slow. Therefore, until drainage occurs, the load is carried by the water in the voids. The load thus causes an increase in the water pressure. The magnitude of the increase, under ideal conditions, is equal to the increment of load. As the water flows from the soil under the induced pressure, the water pressure gradually dissipates and the load increment is correspondingly transferred to the soil particles. The portion of the total stress that is borne by the soil particles, ostensibly through grain-to-grain contacts, is called the effective stress. The relation at any instant is as follows:

$$\sigma_t = \sigma' + \mu \qquad\qquad [2.1]$$

where σ_t is the total applied stress;

   σ' is the effective stress (intergranular pressure); and

   μ is the neutral stress (water pressure).

The significance of the relation is that the response of soil to applied loads is dependent on the effective stress, not the total stress. The water in the voids, although able to sustain a compressive load if flow is restricted, cannot withstand a shearing stress. Shear stresses and strains are the source of failures and, often, of the large displacements that can occur within a soil volume. Effective stresses cannot be measured directly. They can be obtained only by determining the neutral stress (water pressure) and the total stress and calculating the difference.

In some cases, the neutral stress can be estimated from the external conditions. If the soil is saturated and the soil particles form a stable arrangement, the increase in water pressure due to an applied load is equal to the increase in total stress caused by the load. Thus, for this case, the ratio of the increment of neutral stress to increment of total stress is:

$$\delta\mu/\delta\sigma_t = 1 \qquad\qquad [2.2]$$

If the soil is completely dry, there is no increase in water pressure because the applied load is transmitted directly to the soil. The pressure/stress ratio becomes:

$$\delta\mu/\delta\sigma_t = 0 \qquad\qquad [2.3]$$

For a soil with the voids containing both air and water, the ratio of neutral stress to applied total stress will fall between these two values. The actual ratio is dependent on the percentage of the void space filled with water, the stress levels and the soil type. No general relationship among these factors has been determined, but Black and Lee (1973) show that the ratio is no more than 0.1 at 50% of voids filled with water, and at 90% the ratio is about 0.4. The ratio can be determined from laboratory tests on representative samples, but the tests require a relatively high level of skill and special equipment.

When the water content of the soil is below the saturation value, the forces within the soil volume resist further desiccation. These forces of attraction at the soil particle surfaces result in a negative (less than atmospheric) pore water pressure that, in effect, actually adds to the total applied stress so as to increase the effective stress. These negative pressures, often referred to as water suction, greatly complicate analysis in terms of the effective stress.

The difficulty of obtaining the pore pressure/total stress ratio has been the reason that most measurements of the soil response to applied loads has been done in terms of the total stress values. In many instances, this is a satisfactory approach, but it can also result in misleading analyses.

Soil description

Soil exists in essentially infinite variety, but in order to be able to transfer information about soils, researchers have devised a number of classification systems that group soils into a limited number of categories. Groups are formed on the basis of those characteristics that seem most significant to the major interest of the observer. All such systems are limited in their scope and are arbitrary in that there are usually no obvious breaks in the continuum that mark a change from one type of response to another. Furthermore, a system that is useful for one objective is seldom satisfactory for another.

A system that is widely used to group soils according to their mechanical response to stresses and strains is the Unified Soil Classification System (ASTM Standards, 1975). It will be used in this chapter as a means of describing the general soil type under discussion. This system is based on the size and distribution of the soil particles and the effect of water on the consistency of the particles smaller than a specified size. The complete system is laid out in Table 2.1.

The first major separation is to divide all soils on the basis of the size of the particles as determined by a sieve analysis. Those with more than half of the particles larger than 75 μm are called coarse-grained and the others are called fine-grained soils. Coarse-grained soils are further broken down into eight sub-groups on the basis of the sizes of particles and their relative abundance. Fine-grained soils are sub-divided into six sub-groups on the basis of the Atterberg limits, which are measures of the effect of water on the consistence of the soil. Two values are used, the liquid limit and the plastic limit, which determine the amount of water (as a percentage of the weight of the soil particles) necessary to cause a certain change in the consistency of the soil. The amount, in turn, depends on the kind and amount of clay mineral present in the soil sample. In some instances, the amount of water is significantly affected by the presence of decomposed organic material for which case two of the six sub-groups are designated. A completely separate category, termed 'peat', is reserved for soils with large quantities of undecayed fibrous or woody organic material.

Table 2.1. Unified Soil Classification System*

Major divisions			Group symbols	Typical names
Coarse-grained soils — More than 50% retained on No. 200 sieve[†]	Gravels — 50% or more of coarse fraction retained on No. 4 sieve	Clean gravels	GW	Well-graded gravels and gravel–sand mixtures, little or no fines
			GP	Poorly graded gravels and gravel–sand mixtures, little or no fines
		Gravels with fines	GM	Silty gravels, gravel–sand–silt mixtures
			GC	Clayey gravels, gravel–sand–clay mixtures
	Sands — More than 50% of coarse fraction passes No. 4 sieve	Clean sands	SW	Well-graded sands and gravelly sands, little or no fines
			SP	Poorly graded sands and gravelly sands, little or no fines
		Sands with fines	SM	Silty sands, sand–silt mixtures
			SC	Clayey sands, sand–clay mixtures
Fine-grained soils — 50% or more passes No. 200 sieve[†]	Silts and clays — Liquid limit 50% or less		ML	Inorganic silts, very fine sands, rock flour, silty or clayey fine sands
			CL	Inorganic clays of low to medium plasticity, gravelly clays, sandy clays, silty clays, lean clays
			OL	Organic silts and organic silty clays of low plasticity
	Silts and clays — Liquid limit greater than 50%		MH	Inorganic silts, micaceous or diatomaceous fine sands or silts, elastic silts
			CH	Inorganic clays of high plasticity, fat clays
			OH	Organic clays of medium to high plasticity
Highly organic soils			PT	Peat, muck and other highly organic soils

*After ASTM (1982)
[†]Based on the material passing the 75 mm (3 in) sieve

Table 2.1. (contd.)

		Classification criteria	
Classification on basis of percentage of fines	Classification on basis of percentage of fines — GW, GP, SW, SP / GM, GC, SM, SC — Borderline classification requiring use of dual symbols — Less than 5% pass No. 200 sieve / More than 12% pass No. 200 sieve / 5% to 12% pass No. 200 sieve	$C_u = D_{60}/D_{10}$ Greater than 4 $C_c = \dfrac{(D_{30})^2}{D_{10} \times D_{60}}$ Between 1 and 3	
		Not meeting both criteria for GW	
		Atterberg limits plot below 'A' line or plasticity index less than 4	Atterberg limits plotting in hatched area are borderline classifications requiring use of dual symbols
		Atterberg limits plot above 'A' line and plasticity index greater than 7	
		$C_u = D_{60}/D_{10}$ Greater than 6 $C_c = \dfrac{(D_{30})^2}{D_{10} \times D_{60}}$ Between 1 and 3	
		Not meeting both criteria for SW	
		Atterberg limits plot below 'A' line or plasticity index less than 4	Atterberg limits plotting in hatched area are borderline classifications requiring use of dual symbols
		Atterberg limits plot above 'A' line and plasticity index greater than 7	

Plasticity Chart
For classification of fine-grained soils and fine fraction of coarse-grained soils
Atterberg limits plotting in hatched area are borderline classifications requiring use of dual symbols
Equation of A line:
$PI = 0.73(LL - 20)$

Visual–manual identification, see ASTM Designation D2488.

Soil strength

Soil strength concerns the ability of a particular soil to resist the application of a force or a set of forces. It carries with it the implication of a failure or at least a change in the state of the soil as a result of those forces. Strength is the value (or values) that completes the cause–effect relation between a force application to a soil and the final state of that soil. It is a property of a soil not only as a result of the particular combination of particles, but also as a consequence of the particular arrangement of the particles and their conjunction with the water and any other constituents present. Therefore, strength must be measured for each specific soil condition.

The concepts of strength are most readily illustrated in terms of a triaxial compression specimen. In this test, a sample of soil is subjected to an all-round confining pressure and then loaded to failure in axial compression. The triaxial test does not require analysis to get principal stresses as they are the applied stresses. Compressive stress is used because soil cannot sustain tensile stress.

Mohr–Coulomb model

The most frequently used description of soil strength is the one contained in the Mohr–Coulomb model of a soil. One reason for its wide acceptance is that the model considers shear as the limiting condition and it has been observed that soil failures usually occur with a visible shear surface. Another reason is that its use in analysis of failures provides reasonable agreement. The model also is easily visualized in the graphical portrayal of the stress state by the two-(or three-)dimensional Mohr stress circles (Fig. 2.1). In the triaxial test, the axial stress is the major principal stress σ_1, and the confining pressure is the minor principal stress σ_3.

The Mohr–Coulomb model states that a material fails when the shear stress on the failure plane reaches a unique function of the normal stress on that plane. It further states that for different magnitudes of stress, the failure stress circles define an envelope that locates the failure plane as the point of tangency of the envelope with the stress circle. The envelope can be a curved line but, for most soil conditions, a straight line adequately describes the envelope. The relation is

$$\tau_{ff} = c + \sigma_{ff} \tan \varphi \qquad\qquad [2.4]$$

where τ_{ff} is the shear stress on the failure plane at failure;
σ_{ff} is the normal stress on the failure plane at failure;
φ is the slope of the straight line failure envelope; and
c is the intercept of the failure envelope on the shear axis.

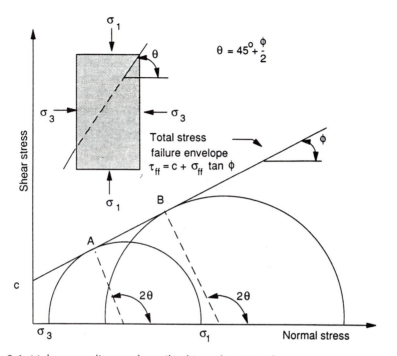

Fig. 2.1. Mohr stress diagram for soil subjected to triaxial compression.

The angle of the failure envelope is called the friction angle, and the term tan φ is the coefficient of friction. This coefficient is the measure of the amount that strength increases as a result of the increase in stress on the failure plane. The intercept c is the cohesive strength of the material and is a constant.

The parameters of the failure envelope for soils are obtained from shear tests on representative samples. To be considerd soil properties, the parameters must be in terms of the effective stresses. If the total stresses are used to determine c and φ, the drainage conditions of the sample during the test affect the values obtained (Holtz and Kovacs, 1981). The convention has been adopted that the effective stress parameters be designated as c' and φ' to distinguish them from the total stress parameters. Total stress parameters can be used in analysis only if the conditions being analysed have the same soil–water–air phase relations and same rate of loading as the test samples used to obtain the parameters.

In terms of effective stresses, coarse-grained soils have zero cohesion at zero effective stress. Fine-grained soils may show a cohesive component if they previously had been consolidated or compacted to a density representative of a pressure greater than the pressure under which the triaxial test is run. It is possible that the cohesion that is indicated is simply due to an

inability to measure accurately the negative pore water pressure that develops
in soils of this nature.

Soils that contain air as well as water in the void space are particularly
difficult to analyse. Such soils are referred to as being 'unsaturated' as
opposed to the soils with all voids filled with water. The air in the voids is
compressible and can for a brief period sustain a pressure due to an applied
load. The water in the voids is actually in tension due to capillary forces and,
more importantly, to the attractive forces of the clay mineral surfaces. This
special case will be discussed further in a subsequent section.

It just be noted that the Mohr–Coulomb model is concerned only with
the maximum shear stress at failure. The stress–strain characteristics of the
material tested do not enter the analysis.

Critical state model

The critical state model of the response of soil to stress was developed by
Roscoe *et al.* (1958) and Schofield and Wroth (1968). The critical state is
defined as the condition of the soil for which shear strain will cause no increase
in stress or change of the volume occupied by the soil. The soil flows like
a frictional fluid. At the critical state, there is a unique relation among the
parameters that represent the soil, the stress and the soil volume. The
conditions assumed for the evaluation are that the soil is saturated and
subjected to a triaxial state of stress.

The critical state is defined by the conditions:

$$q = Mp'   \qquad [2.5]$$

and,

$$v = \Gamma - \mu \ln p'   \qquad [2.6]$$

where q is the principal stress difference, $\sigma'_1 - \sigma'_3$;
        M and μ are material constants;
        p' is the mean effective stress, $[(\sigma_1 + \sigma_2 + \sigma_3)/3] - u$,
        where u is the pore pressure;
        v is the specific volume of the soil, $(1 - e)$,
        where e is the void ratio; and
        Γ is the specific volume at $p' = 1\,\text{kPa}$.

The critical state concept is portrayed graphically in Fig. 2.2. A soil is viewed
in the three-dimensional q–p'–v space. The soil can exist within the bounds
of three limiting surfaces. The tensile surface defines the limitation that soil
cannot sustain a tensile stress. The Hvorslev surface recognizes that the
interaction of this surface with any q–p' plane is a line analogous to the
failure envelope of a Hvorslev (1937, 1960) failure model. It expresses the
dependence of the principal stress difference on the mean effective stress.

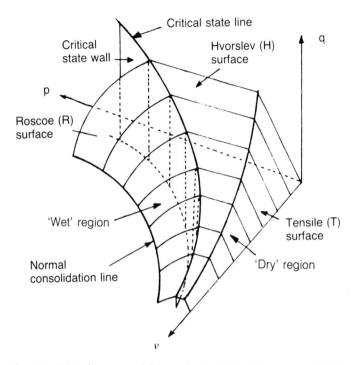

Fig. 2.2. Critical state model boundaries. (After Kawamura, 1985.)

The critical state parameter, M, is related to the parameter tan φ. The cohesion intercept on the tensile surface decreases as the specific volume increases (soil density decreases). The Roscoe surface, named after a founder of the theory (Roscoe *et al.*, 1958), denotes the occurrence, in some instances, of plastic flow instead of a shear surface. The intersection of the Roscoe surface with a $p'-v$ plane is analogous to the void ratio versus log pressure curve of soil consolidation. The critical state parameter, μ, is similar to the compression index.

The critical state line is the intersection of two limiting surfaces; the R and H surfaces as shown in the figure. An element of soil originally with specific volume v_0 and mean stress p' when stressed additionally to acquire a stress difference q, will deform and experience change in both p' and v. The soil can be in any state that lies beneath the limiting surfaces; if the combination of parameters bring it to a surface it moves along the surface until it reaches the critical state line. After that q and v can no longer change and the soil distorts without change of these parameters. If the soil first reaches the H surface, it will fail with well-defined shear surfaces. If it first reaches the R surface, it will fail plastically.

The parameters for the critical state model must be obtained in terms of effective stresses from tests of representative samples. In addition, the volume changes must be measured. Thus application of the critical state model to unsaturated soil presents all of the difficulties of the other models.

Interaction equations

Interaction equations are intended to relate the properties of a material, the soil, to the geometry and kinematics of a tool to provide a prediction of the forces on the tool and the motion of the soil. They express the mechanics of a soil–machine operation. The requirements for a complete mechanics of a complex machine are so great that it is not feasible at this time to evaluate a complete system. Instead, the strategy has been to deal with components of the total and to endeavour to link them together empirically. Analytic solutions to several basic interactions have been obtained and have been used with varying degrees of success. The bearing capacity of a soil – that is, its ability to support a vertically applied load – is the oldest and most widely accepted of these. The passive resistance of soil to a very wide plate cutting a thin slice from the surface has also been used extensively, most recently in the form of the 'universal earthmoving equation'. Needed, but not yet established in this category, is the force analysis of a narrow tine drawn through the soil and an analytical mechanics of wheel traction for at least a simple basic case. For the latter two cases, there have been several proposed solutions, but none has attained wide acceptance.

Bearing capacity

The ultimate bearing capacity of a loaded area depends on the shear strength of the supporting soil. The equations that describe the relation between the soil strength and the maximum load were developed by Terzaghi (1943). The analysis considers an infinitely long, narrow footing resting at or near the surface of a soil that is characterized by Mohr–Coulomb parameters. Under increasing load, potential shear surfaces in the form of logarithmic spirals begin to form in the soil beneath the loaded area (Fig. 2.3). At failure, the footing will abruptly sink, often rotating along one of the spiral surfaces. The analysis procedure sums the support received by the footing from the soil parameters of cohesion, friction and weight. The resulting expression is:

$$q_u = cN_c + qN_q + \frac{1}{2}\gamma BN_\gamma \qquad [2.7]$$

where q_u is the ultimate bearing capacity in force per unit area;
c is the soil cohesion;
q is the surcharge pressure above the base, usually simply

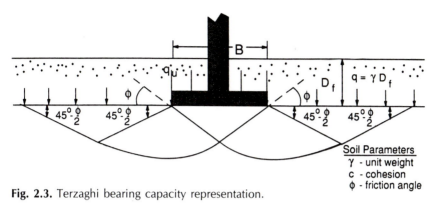

Fig. 2.3. Terzaghi bearing capacity representation.

Soil Parameters
γ - unit weight
c - cohesion
φ - friction angle

$q = \gamma D_f$ where D_f is the depth to the base of the footing;
γ is the unit weight of the soil;
B is the width of the footing; and
N_c, N_q and $N\gamma$ are dimensionless bearing capacity factors
that are functions of the soil friction angle.
Possible solutions for the bearing capacity factors were summarized by Vesic (1973) with the following recommendations:

$$N_q = e^{\pi \tan \varphi} \tan^2 \frac{45° + \varphi}{2} \qquad [2.8]$$

$$N_c = (N_q - 1) \cot \varphi \qquad [2.9]$$

$$N_\gamma = 2(N_q + 1) \tan \varphi \qquad [2.10]$$

The bearing capacity equations were developed for a long footing at or near the soil surface. Theoretical analysis of footings of other shapes and depths of embedment has not been accomplished, but empirical factors have been proposed to adjust for these situations (Vesic, 1973).

Civil engineers have used these equations to analyse the adequacy of shallow foundations and have found them satisfactory. It is of interest to note, however, that in most of the validation tests in fine-grained soils, the soil properties were stated in terms of the total stresses. The justification has been that for practical loading times, the pore water pressures that were caused by the applied load did not have time to dissipate and therefore the total stress condition was the critical one.

Earthmoving equation

Reece (1965) showed that the problem of the force required to move a plate horizontally into a soil surface (Fig. 2.4) could be approached in a manner

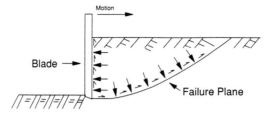

Fig. 2.4. Soil failure pattern ahead of a wide cutting blade, assuming internal friction.

similar to the bearing capacity problem. He derived an expression for the horizontal force on a wide blade by using the same concept of adding the resistance components due to cohesion, friction, soil weight and the adhesion that develops between the metal blade, as follows:

$$P = czbK_c + qzbK_q - \gamma z^2 bK_y + c_a zbK_a \qquad [2.11]$$

where P is the total force on the blade;
        c is the soil cohesion;
        q is the surcharge on the failure surface;
        γ is the unit weight of the soil;
        c_a is the adhesion of the soil on the metal blade;
        z is the depth of cutting;
        b is the width of the blade; and
        K_c, K_q, K_y and K_a are the dimensionless earthmoving numbers.
Reece (1965) and others use the symbol N as in the bearing capacity equations, but K is used here because the values are quite different. The values of the K numbers depend on the soil friction angle, soil–metal adhesion, and the angle of inclination of the blade to the soil surface. A tabulation of the K values is given in Hettiaratchi and Reece, 1974, and reproduced in McKyes, 1985. The numbers are valid for blades that are wide relative to the depth of cut (b/z at least 4:1). For narrow blades, the soil flow at the sides contributes significantly to the total force and requires a different calculation.

Corroborating data to support the validity of the analysis have been largely confined to soil bin data and have been based on the Mohr–Coulomb model using total stresses.

Dimensional analysis

The principles of dimensional analysis and model testing have been applied to soil–machine systems by numerous researchers and engineers. The technique permits the characterization of complex systems which are not completely understood. The theory is based upon the concept of dimensionally homogeneous equations. The form of such equations is independent of the fundamental units used; that is, they do not contain dimensional constants.

The advantage of the technique is that a complicated system can be reduced to a set of dimensionless constants involving the principal variables which can then be quantitatively related to each other by experiment. Thus, a usable predictive system can be developed without the depth of understanding of the physical phenomena that is required for development of closed-form mathematical prediction equations. The studies of soil–machine interactions for chisel ploughs by Reaves *et al.* (1966), pneumatic tyres by Freitag (1965), and earthmovers by Sullivan (1964), are examples of dimensional-analysis-based predictive and scale-modelling systems.

Finite element method and rheology

Yong and Fattah (1976), Yong and Hanna (1977), Yong *et al.* (1978) and Perumpral and Desai (1979), among others, have applied the finite element method (FEM) to strain and displacement states of soil–machine systems. Results indicate that the finite element method is useful in soil dynamics, although there is come concern over applications involving large strains and high strain rates. FEM, if successful, would make possible *generalized* soil machine models that would greatly broaden the scope of our predictive/design capability. It also may contribute to our rationalization and measurement of soil properties significant to soil–machine systems. Even a moderate success in either one of these potential applications would make FEM of great value to the field of soil dynamics.

Tillage

The application of soil mechanics principles to the analysis and prediction of the forces on tillage tools has evolved from codified experience to relatively sophisticated mathematical analyses. The set of empirical relations developed by the American Society of Agricultural Engineers (ASAE, 1990) are representative of the most elementary level. The ASAE system was developed by a number of collaborating engineers and represents the distillation of a great deal of field and laboratory experience. It has two basic components: a power function relating specific resistance values to speed for 'typical' soil types, and a table of a range of draught requirements for a wide variety of tillage tools (Table 2.2). While these relations permit useful estimates to be made of tillage forces, the level of precision is not sufficient for many applications, particularly those involving unusual soil conditions.

The development of a more sophisticated approach has been slowed by the combined complexity of the soil and the tools. The properties of the soil vary with water content and degree of compaction, while tillage tools are made in a variety of configurations. There have been many studies that relate

Table 2.2. Typical tillage forces.

Draught vs. speed

Mouldboard plough

$R = 6 + 0.053\ S^2$	Decatur clay loam
$R = 4.8 + 0.024\ S^2$	Silty clay loam
$R = 2 + 0.013\ S^2$	Sand

where R is draught per unit area of cross section of furrow in N/cm^2 and S is speed in km/h.

Chisel plough

$D = 520 + 49.2\ S$	Loam
$D = 480 + 48.1\ S$	Clay loam
$D = 520 + 36.1\ S$	Clay

where D is draught in N per tool spaced at 30 cm operating in firm soil at depth of 8.26 cm and speed S in km/h.

Typical draught forces for implements independent of speed

Implement	Draught
Spike tooth harrow	440–730 N/m width
Rod weeder	880–1830 N/m width
Roller/Packer	440–880 N/m width
Ammonia applicator	1800 N per knife
Row planters	450–800 N/row
Grain drill	130–670 N/opener

Source: After data from ASAE, 1990.

the geometrics of the soil–tool arrangement to the forces on the tool during tillage. This includes such factors as the tool dimensions, rake angle, the depth of cut and the shape of the tool surface. There have also been studies that show that the magnitude of the force on a tool of a particular shape can be related to a measure of the strength of the soil.

The effect of tilling speed on resulting machine forces is another complicating factor in predicting tillage tool performance. The higher the speed, the greater the force required to maintain that speed. Part of the increase is due to inertial effects related to moving and accelerating the mass of soil involved. The remainder of the increase is the result of an effective increase in the shear strength of the soil. Wismer and Luth (1971) in a study of the forces on a simple blade, demonstrated that inertial forces accounted for all of the increase in draught force in dry sand. Tests in a saturated clay showed that inertial effects could account for only about half of the measured increase in force. Cohesive, damp, frictional soil displayed an intermediate behaviour.

A classical mechanics approach to the analysis of the tillage tool–soil interaction has been utilized by a number of investigators (Zelenin, 1950; Payne, 1956; Söhne, 1956; Osman, 1964; Reece, 1965; Hettiaratchi *et al.*, 1966; Hettiaratchi and Reece, 1967; Godwin and Spoor, 1977; McKyes and Ali, 1977; Grisso *et al.*, 1980; Perumpral *et al.*, 1983). All have employed the Mohr–Coulomb model of soil strength and the concept of passive soil resistance (Fig. 2.4). None has tried to incorporate directly the effects of the speed at which the tool was moving. The analyses usually are restricted to certain tool shapes; most commonly either a wide flat blade or a narrow one (a tine). All result in closed-form solutions, but several depend on such simplifying assumptions as straight line failure surfaces and idealized frictional relations on the tool surfaces.

Plasse *et al.* (1985) and Grisso and Perumpral (1985) each analysed four variants of analytical procedures for narrow flat blades and compared the results with the experimental data available from a number of sources in the literature. All variants followed the pattern of additive terms employing the dimensionless coefficients similar to the approach used by Reece (1965). The primary difference among the procedures was the manner in which the flow of soil around the sides of the blade was treated and whether a curved or plane failure surface was assumed. Soil properties were those measured by the initial authors; or, where none was provided, estimates were made from soil descriptions. The Hettiaratchi and Reece (1967) procedure clearly overestimated the horizontal forces but the remainder, those by Godwin and Spoor (1977), McKyes and Ali (1977), Grisso *et al.* (1980), and Perumpral

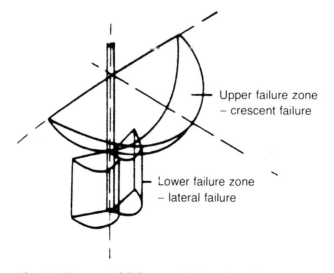

Upper failure zone
– crescent failure

Lower failure zone
– lateral failure

Fig. 2.5. Conceptual failure mechanism for a deep narrow tool. (After Godwin and Spoor, 1977.)

et al. (1983), grouped fairly closely around the test data. The Godwin and Spoor procedure was as good as any and is representative of the concepts involved; it has been chosen to represent the state-of-the-art of tillage tool analysis in the remainder of this section.

Godwin and Spoor, using glass-sided box observations of soil patterns formed with deep tines, concluded that the failure mechanism for these tools involved the formation of a compacted soil wedge which below a critical depth moved laterally, while above this depth the soil moved along in a crescent failure surface which moved upward (Fig. 2.5). Using the principles of Mohr–Coulomb failure and the universal earthmoving equation with some simplifying assumptions of the shape of the failure surfaces produced, force response equations were developed for deep tines. Estimates of tine forces require definition of the critical depth, which can be accomplished either by iterative techniques or by differentiation.

While narrow blades are representative of a large class of tillage tools, there remain many that are not understood well enough to have developed a comparable level of analytical capability. There remains much to challenge a new generation of researchers.

Earthmoving

The development of a mechanics of earthmoving has followed a similar path as that for tillage although at a much lower level of effort, except for work in the USSR. The design procedures used by equipment manufacturers usually are not made public. The data provided by a major manufacturer in its users' handbook are based on general descriptions of soil types and range values similar to that given by the ASAE for tillage tools (Nichols, 1976; John Deere, 1980).

Freitag (1988) summarized much of the recent Soviet soil cutting and excavation research, and included an extensive bibliography. Since the interactions between soils and earthmoving blades is very complex, much of the work to date has been of an experimental rather than a theoretical nature. Zelenin *et al.* (1975) offer a number of design rules for earthmoving tools involving geometry, the use and spacing of cutting teeth, and the like. Caterpillar has pioneered the application of model-testing techniques to equipment development (Cobb *et al.*, 1961; Sullivan, 1964) and dimensional analysis is frequently used. A common set of dimensionless numerics and scale factors to assist in design and analysis studies is given in Table 2.3. General Motors also contributed to scale-model developments in earthmoving in the early 1960s (Emori and Schuring, 1965). Progress in this area appears to have slowed in the 1970s and 1980s.

Table 2.3. Dimensionless numerics and scale factors for an earthmoving scraper.

Dimensionless Numerics			
$\dfrac{P}{W}$	$\dfrac{P}{\gamma L^3}$	$\dfrac{c}{\gamma L}$	ϕ
$\dfrac{v^2}{Lg}$	$\dfrac{L}{d}$	$\dfrac{vt}{L}$	f

Derived scale factors			
Symbol	Quantity	Scale factor*	Scale for 1/10 scale model
W	Weight of excavated soil	λ^3	1/1000
P	Horizontal pushing force	λ^3	1/1000
L	Linear dimension specifying the size of the bowl	λ	1/10
v	Forward velocity of the bowl	$\lambda^{\frac{1}{2}}$	$\sqrt{1/10}$
d	Cutting depth	λ	1/10
t	Cutting time	$\lambda^{\frac{1}{2}}$	$\sqrt{1/10}$
c	Cohesion of the soil	λ	1/10
ϕ	Angle of internal friction of the soil	1	1
f	Coefficient of friction of soil on metal	1	1
γ	Weight bulk density of the soil	1	1
g	Acceleration of gravity	1	1

Source: after Sullivan, 1964.

$$*\lambda = \frac{\text{characteristic length of model}}{\text{characteristic length of prototype}}$$

Hettiaratchi *et al.* (1966) and Palmer *et al.* (1979) developed equations for earthmoving operations based on the universal earthmoving equation. Hettiaratchi's equations appear to be the best available for wide blades at shallow depths, while Godwin and Spoor (1977) appear best for narrow, deep tools. Reece demonstrated the application of these equations to the geometric design of a cable-laying sea plough.

The soil–machine mechanics of earthmoving appears to be getting little research effort considering its commercial scope in modern society. This could be due to the maturity of the industry, or because it is dominated by commercial firms that do not publish research efforts or results widely. In any event, with the increase in tunnelling construction, dredging and island building, it would seem to be a good area for research emphasis.

Traction

Of all the areas of soil–machine mechanics, traction has historically received the greatest research emphasis. The application of traction mechanics has progressed from the relatively simple semi-quantitative empirical level of the 1960s to the computer based mobility modelling systems of the 1990s.

The US Army Corps of Engineers Waterways Experiment Station (WES) performed many trafficability studies beginning in the 1940s. This work was directed at developing a fifty pass 'go, no-go' criterion for military vehicles due to their requirement for column formation. The procedure consisted of determining the soil strength level, using a cone penetration test, at which a given vehicle could just complete fifty passes in its own tracks. This soil strength was defined as the 'vehicle cone index'. After a sufficient number of tests had been run, this value could be computed for a real or hypothetical vehicle based upon its running gear characteristics such as number of tyres, weight, ground clearance, and so on as reported by Knight (1956). An added dimension of this technique was the ability to predict additional drawbar pull or slope climbing capability that would result in soil conditions exceeding the vehicle cone index by a specified amount (Fig. 2.6). The cone penetrometer test developed by WES is a bearing capacity test using a cone probe (ASAE, 1989). The variability of naturally occurring soil requires a large number of individual soil measurements if adequate definition of the strength affecting machine performance is to be determined. The cone penetrometer is a very quick test that measures strength variations with depth, which also greatly affects machine performance. This makes it attractive for use in traction analysis.

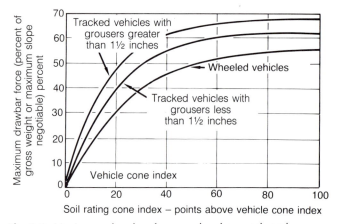

Fig. 2.6. Maximum drawbar force on level ground, and maximum slope that can be climbed, for self-propelled vehicles on fine, drained soils and sands with fines, poorly drained. (After Knight, 1956.)

Söhne (1963) defined the practical limits of traction and motion resistance of wheeled vehicles on a variety of terrains. Zoz (1970, 1987) extended Söhne's tractive performance curves to include efficiency concepts and this was incorporated in the ASAE Data Standard D497. These relations couple tractor and implement specifications and can be used to predict the performance of existing tractors in untested conditions or conceptual tractors in a range of conditions. In addition, the relations can be used to ensure proper matching of tractor and implement and to optimize tractor power, ballast, and tyre size.

Nuttall (1956) applied dimensional analysis to soil–wheel systems in an attempt to establish methods for predicting the performance of full-size vehicles. This work resulted in the identification of the principal dimensionless numerics governing soil–vehicle performance and experimental techniques for testing model and prototype soil–vehicle systems. Freitag (1965) extended this approach to the development of traction numerics from the results of extensive soil bin tests of pneumatic tyres operating in air-dry sand and saturated fine-grained soils. Cone index determined from cone penetrometer tests was used as the measure of soil strength in these relations. The clay numeric developed was applicable to water-saturated, completely remoulded clays which exhibit a zero angle of internal friction. The clay numeric as originally formulated was:

$$\frac{\mathrm{CI}\, bd}{W} \left(\frac{\delta}{h}\right)^{0.5} \tag{2.12}$$

where CI = average cone index in the 0 to 150 mm layer;
  b = unloaded tyre section width;
  d = unloaded tyre diameter;
  h = unloaded tyre section height;
  δ = tyre deflection;
  W = tyre load.

This numeric defined the soil, tyre geometry and loading variables and was uniquely related to the tractive performance numerics:

$$\text{Pull number} \left(\frac{P_{20}}{W}\right) = \frac{\text{tyre pull at 20\% slip}}{\text{tyre load}} \tag{2.13}$$

$$\text{Towed force number} \left(\frac{P_{\mathrm{T}}}{W}\right) = \frac{\text{towed force}}{\text{tyre load}} \tag{2.14}$$

$$\text{Sinkage number} \left(\frac{z}{d}\right) = \frac{\text{sinkage of tyre}}{\text{tyre diameter}} \tag{2.15}$$

This work was extended by Rula and Nuttall (1971) and Turnage (1972) to include additional field soil variables and a broader range of tyre size and shape. Rating cone index, a modified version of cone index accounting for the loss of soil strength experienced under vehicular traffic, replaced simple cone index in these relations (Fig. 2.7). These relations are valid for cohesive soils at the towed and 20% slip conditions. These relations have been accepted by the NATO military forces, and are incorporated in the reference model for evaluating new military vehicle designs and the determination of tactical mobility of existing vehicles in a given theatre of operation (NATO reference Mobility Model, 1979).

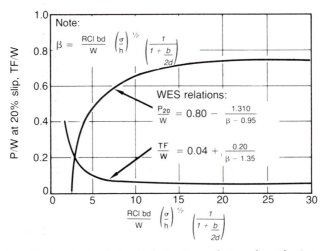

Fig. 2.7. WES wheeled vehicle traction relations for cohesive soils. (After Rula and Knight, 1971; Turnage, 1972.)

This WES approach was applied by Dwyer *et al.* (1975) to agricultural machinery use. Field experiments in a variety of agricultural field conditions using a single-wheel tester and a wide range of agricultural tyre sizes produced tyre–soil correlations for agricultural use. These relations have been incorporated in a tyre selection, tractor ballasting and implement tyre sizing guide for farmers in the United Kingdom.

General off-road vehicle simulation programmes require continuous mathematical relations between the soil–wheel numeric and the performance numerics for all slip values and for cohesive-frictional soils. Wismer and Luth (1972) extended the WES approach to include slip and cohesive-frictional soil factors. The resulting equation is based upon a net traction (pull/weight) relation as follows:

$$\frac{P}{W} = 0.75\,(1 - e^{-0.3C_n S}) - \left(\frac{1.2}{C_n} - 0.4\right) \qquad [2.16]$$

where P = wheel pull, parallel to soil surface;

W = dynamic wheel load, normal to soil surface;

S = wheel slip;

e = base of natural logarithms;

$$C_n = \frac{CI\,bd}{W};$$

CI, b and d as before.

The pull/weight relation is accompanied by a definition of tractive efficiency, ratio of output power over input power of the drive wheels, as follows:

$$\text{TE} = \frac{\text{output power}}{\text{input power}} = \left\{ 1 - \left[\frac{(1.2/C_n) + 0.4}{0.75(1 - e^{-0.3C_nS})} \right] \right\} (1 - S) \qquad [2.17]$$

These relations are shown graphically in Fig. 2.8 and are considered applicable to agricultural, earthmoving and forestry vehicle–soil systems using pneumatic tyres of an approximate b/d ratio of 0.3, a tyre deflection/section height ratio of approximately 0.2 and a tyre radius/diameter ratio of 0.475. These equations have been used in a number of vehicle simulation/analysis programmes, and have been incorporated in the ASAE Agricultural Machinery and Management Data Standard (ASAE D497). Brixius (1987) extended and modified these relations incorporating more tyre geometry factors.

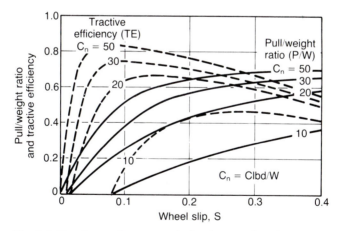

Fig. 2.8. Tractive performance of wheels on soil. (After Wismer and Luth, 1972.)

Bekker (1956) compiled the first complete engineering model of soil–vehicle traction based upon classical mechanics. It is applicable to both wheeled and tracked vehicles and employs six soil values obtained from rotating ring shear and plate penetration tests. Net traction is predicted as the difference between gross thrust and motion resistance. The Bekker equations define the gross tractive effort of a wheel or track as follows:

$$H = (A_c + W \tan \phi) \left[1 - \frac{K}{Sl} \left(1 - e^{\frac{-Sl}{K}} \right) \right]$$ [2.18]

where W = load on tractive element;
        S = slip ratio = 1 − (actual velocity/theoretical velocity);
        c = soil cohesion from ring shear test;
        ϕ = soil internal friction angle from ring shear test;
        l = length of contact area = $\sqrt{\delta(d - \delta)}$;
        δ = tyre deflection;
        d = tyre diameter;
        K = tangent modulus of deformation from ring shear test;
        A_c = area of contact between tyre and soil.

Motion resistance of a wheel or track is predicted from a modified Bernstein (1913) equation incorporating the concept that the sinkage exponent n was a soil property independent of plate geometry:

$$p = \left(\frac{k_c}{b} + k_\phi \right) z^n$$ [2.19]

where k_c = cohesive soil modulus from plate penetration;
        k_ϕ = frictional soil modulus from plate penetration;
        b = width of plate or tyre contact area;
        n = sinkage exponent;
        z = depth of plate penetration.

Experience with this model in the 1960s indicated that the soil−plate constants k_c, k_ϕ and n were not independent of soil type or plate size. Reece (1964) suggested a modified version of the Bekker compaction resistance equation:

$$p = \left(ck'_c - \gamma \frac{b}{2} k'_\phi \right) \left(\frac{z}{b} \right)^n$$ [2.20]

where k'_c, k'_ϕ, and n are modified values of the original soil−plate fitting parameters, c is soil cohesion, and γ is soil unit weight.

    In addition to Reece, Janosi (1961), Pavlics (1966), Karafiath and Nowatzki (1978), Wong (1989) and others have applied the Bekker equations to engineering problems and recommended modification. The most current, state-of-the-art, description of the modified Bekker equations is presented by J.Y. Wong (1989).

Compaction

Compaction implies dynamically forcing the particles in a soil into a smaller space and, in the process, changing the soil fabric (the spatial relations among the particles). The reduction in volume occupied by the soil is accomplished quickly, usually in a fraction of a second, by expulsion of air from the void space. Compaction is caused by transient surface loads applied by machines or any other traffic over the surface.

If a soil compacts below a critical porosity, plant growth and productivity decline. The problem is twofold; first, plants require a minimum amount of air and void space to do well and, second, the soil must not be so hard that the roots are not able to penetrate. The amount of water present can greatly affect the influence of soil compaction. At any particular porosity, more water in the voids reduces the hardness of the soil, but at the same time displaces air that may be needed by the plants. Thus, crop production can either increase or decrease with increasing amounts of soil compaction depending upon conditions. Soil that is too loose may result in poor germination, high evaporation loses and limited capillary movement of water. Figure 2.9 shows this effect in terms of the relative yield of a field in relation to soil compactness for dry, normal and wet crop years.

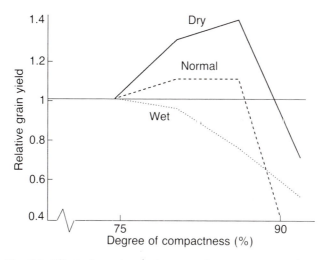

Fig. 2.9. Effect of weather and compaction on crop growth. (After Voorhees, 1986.)

Experience has shown that compaction problems are most likely in soils that have a wide range of grain sizes and soils with large amounts of clay or organic colloids. Compaction is most likely to occur when the soil moisture is near field capacity. The degree of compaction in the near-surface layers is related to the contact pressure exerted by the traffic. Total loads greater than about ten tonnes have been identified as the cause of compaction at depths below the usual tillage zone. Studies have shown also that 80–90% of the total possible compaction will occur on the first application of a load.

The application of any force to a soil will usually result in some reduction in volume as particles seek new positions of equilibrium under the changed force system. However, in an assemblage of particles, there is a tendency for arrangements to be formed to resist a single sustained force. In Fig. 2.10, the group of particles are in a position to resist densification by the vertically applied force F_v. If, instead, the force is rotated to become F_h, the ability to resist densification is much less. Thus, if force applications are distributed so as to occur successively from many different directions, the amount of densification that will occur is much greater than if a force of the same magnitude were simply maintained. To obtain the maximum densification, the applied forces must overcome the cohesive and frictional resistance between particles in all possible directions, to cause the particles to slide, roll and rotate to provide the greatest probability of finding the densest packing arrangement. The process cannot be expected to occur in one or even a few applications of stress; it is a process of progressively diminishing return for the energy input. If the objective is to minimize densification, the opposite strategy is appropriate.

The stresses that are developed in the soil beneath the wheel of a machine can be estimated reasonably well by the bearing capacity equation (Söhne, 1953; Foster and Ahlvin, 1954; Vanden Berg et al., 1957). The data show that shear stresses, which are $\tau = \frac{1}{2}(\sigma_1 - \sigma_3)$, where σ_1, σ_3 are principal

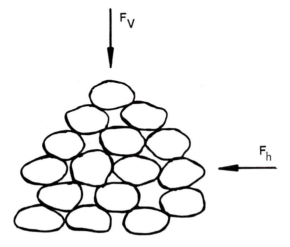

Fig. 2.10. Effect of soil particle arrangement on compaction.

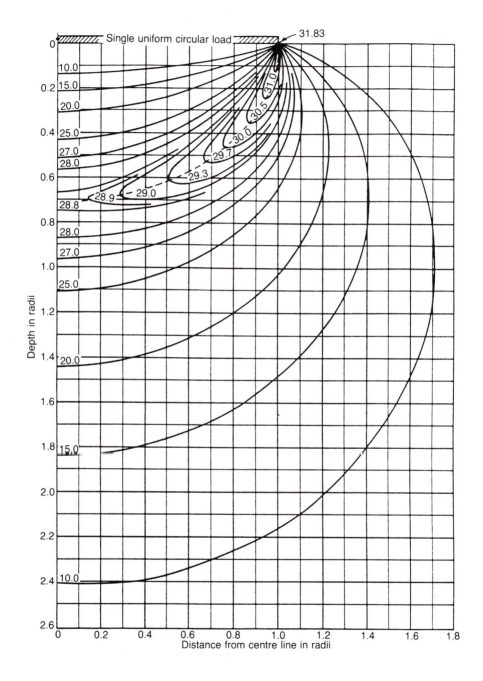

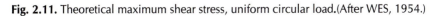

Fig. 2.11. Theoretical maximum shear stress, uniform circular load. (After WES, 1954.)

perpendicular stress components, are most intense at the edge of the applied load (Fig. 2.11). The stresses diminish with depth, especially after a depth equal to about half the width of the loaded area. At a depth of about three times the width of the loaded area, the shear stress intensity is only one-fourth of the maximum. This shows that the effectiveness of a compactor is greatest in a zone just below the contact surface.

Compaction involves both stress and deformation, which suggests that the mechanics of the process is dependent on the work or energy input. Studies by Proctor (1947), Vanden Berg (1966) and Hodek (1972) provide data that support this hypothesis. However, other studies, such as Chancellor and Korayem (1965) and Bailey *et al.* (1985), have derived correlations without using an energy relation. At the time of writing the independent parameters in a compaction mechanics equation appear not to have been established.

The compaction process is relatively easy to visualize for coarse-grained soils, but fine-grained soils are more complex. Because of the active clay component, compaction of these soils is greatly dependent on the amount of water present. For a given level of compaction energy, the porosity of samples compacted at successively higher water contents is found to be less and less until a minimum is reached (Fig. 2.12). The effect of the increasing amounts of water has been to allow the energy input to be more successful in forcing the soil into a closer state of packing.

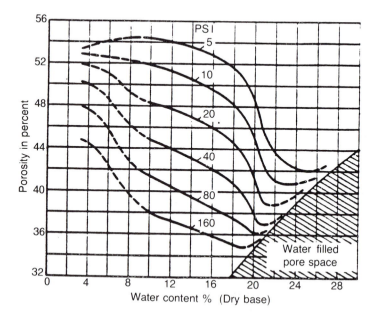

Fig. 2.12. Effect of water content on soil density resulting from different stresses. (After Söhne, 1958.)

Soils that contain clay minerals develop a blocky structure as a result of desiccation. The individual particles cohere in tightly bound packets and the packets behave much like single particles. Within the packets the voids are very fine and may remain saturated with water. Between the packets there are large voids that normally contain little or no water. When the soil is rewetted, the water initially enters only the large voids between the packets. The voids within the packets fill very slowly, if at all, so that equilibrium of water potentials does not occur during a seasonal cycle. The addition of water to such a soil causes the packets to soften at their outer surface. During compaction, the soil first acts like a coarse-grained soil but, if the compaction forces are great enough, the packets deform to fill the adjacent void space. This process has been analysed in some detail by Hodek (1972). At some point, usually about 80–90% saturation, the water so nearly fills the voids that the air exists as trapped bubbles. Then compaction ceases and further application of energy results only in resilient or plastic deformation of the soil mass.

Issues to be resolved

Development of a mechanics of unsaturated soils

Many of the instances of the interaction of working machines with the soil occur when the pore spaces contain some water, but are not completely filled. These include most tillage operations, earthmoving for construction, and any overland operation that seeks to maximize performance and minimize the damage to the terrain. Good engineering solutions to the problems associated with these functions require a sound understanding of the mechanical behaviour of unsaturated soils. To date, some progress has been made by adapting solutions developed for saturated soils which usually only approximate the real conditions.

Much of the successful application of the principal strength models of soil in civil engineering has been achieved in saturated soil and by using the principles of effective stress. Effective stresses explain the immediate response of the soil to loads and allow predictions of the long-term behaviour. It is logical to conclude that if the effective stresses acting on an unsaturated soil are known, the same degree of predictability could be realized for this soil condition. At the same time, however, it must be recognized that the equations that have been described in the previous sections are based on measurements of the total stress, and the soil properties used in the analyses are also in terms of total stresses. Thus the question must be addressed as to whether effective stresses are a necessary factor in the study of unsaturated soil.

The difficulties are encountered primarily with fine-grained soil and arise from the ability of the clay minerals to attract and hold water. The particles in a fine-grained soil are seldom uniformly distributed (Croney et al. 1958). The very small clay particles exist in clusters or packets that group around the larger silt and sand grains. The air voids within the clay packets are relatively small, while the voids between the packets and the silt and sand grains are large. The voids within the packets remain saturated to very high suction values, but the others can empty at low suctions (Brackley, 1975).

In the literature, soils that are not saturated are sometimes referred to as partly saturated and at other times as unsaturated. From a semantic point of view, the latter term is to be preferred as something either is saturated or it is not. However, since it is possible for the clay packets to be saturated at the same time that the large voids are empty, the expression 'partly saturated' may actually convey more meaning.

Aitchison and Donald (1956) and Croney et al. (1958) showed that the moisture tension in an unsaturated soil was related to the increase in soil strength above that attributable to the external stress state. Bishop (1959) proposed that the effective stress relation for an unsaturated soil could be expressed as:

$$\sigma' = \sigma - u_a = \psi(u_a - u_w) \qquad [2.21]$$

where σ is the total stress;
       u_a is the pore air pressure;
       u_w is the pore water pressure; and
       ψ is a factor related to the degree of saturation.

Bishop et al. (1960) and Bishop and Blight (1963) were able to demonstrate that the shear strength of unsaturated soils could be analysed by use of this expression.

Bishop et al. (1960) noted that the expression for the volume change of such soils would at least require a different constant. Jennings and Burland (1962) demonstrated that there were indeed problems in extending the Bishop equation to describe volume changes as a result of shear in unsaturated soil. When the air–water interface has retreated into the smaller pore spaces, the suction no longer acts over the whole cross-section of the soil and the equation does not account for that. Burland (1965) also points out that the pressures due to the air–water tension produce only normal stresses at the points of soil particle contact. An externally applied stress would produce both normal and shear stresses so the two conditions cannot be interchangeable in the analysis of volume changes. It was demonstrated that the outcome of the application of a set of boundary conditions was dependent on the soil structure and the stress path.

Toll (1990) presents an analysis of unsaturated soil in terms of the concepts of the critical state approach. He found it necessary to employ five parameters: $(p - u_a)$, $(u_a - u_w)$, q, v, and S. If the pore air pressure is atmospheric (u_a), the terms are simply:

p the principal stress invariant in terms of total stresses;
u_w The suction of soil moisture tension;
q the principal stress difference;
v the specific volume; and
S the degree of saturation.

In his experimental analysis, Toll found that the effect of the suction declined rapidly with the degree of saturation, and at a degree of saturation less than 50% it had little effect on either the strength or the volume changes. This suggests that some of the usefulness of the existing total stress equations can be attributed to their application to cases for which only the total stress was a factor. Similarly, the cases of poor results may represent a combination of conditions in which the suction still played an important role. It must be noted also that, except for samples that were nearly saturated, the original soil structure had an effect.

The effective stress analysis seems to require such complex and detailed measurements of the stresses and the soil characteristics that it will not be a practical method of field application. However, it also is likely that if the total implication of the actual effective stresses were understood, a simpler approximate total stress approach could provide improved accuracy.

Rate effects

As cited previously, speed can have a significant effect on soil–machine response. The inertial effects can be readily accounted for. Soil strain-rate effects are not as well understood, but can be significant if speed varies over a range of ten or more in fine-grained soils. Current soil dynamics practices do not use strain-rate corrected strength measures. Research efforts focusing on the practical means for measuring and accounting for strain-rate effects on soil–machine response are warranted.

Testing

Progress in science requires testing. This is also true for the engineering sciences of agriculture. Much of the progress in traction mechanics since World War II can be ascribed to government-funded research on traction and the associated test facilities that were developed to support that research. The development of theory without experimental observations to confirm or refute an hypothesis produces little progress in scientific understanding or engineering application. Similarly, the random collection of data unsupported by a unifying theory or hypothesis is grossly inefficient. The field of soil–machine systems would benefit from increased cooperation among researchers and institutions with testing facilities. Common agreement on test methods, or at least better documentation of test procedures, would greatly aid in the usability of test data by others. Government support of tests specifically designed to compare the accuracy of competing theories in a particular area, such as narrow tines, should be considered. But the greatest progress might be made by the adoption of a philosophy of independent verification of the test results of others. The field of soil–machine systems has now progressed to a sufficiently mature state, as witnessed by the preceding paragraphs, that a consolidation of theory and predictive systems would be beneficial. Collaboration among researchers and institutes to independently repeat critical correlation/verification tests could accelerate understanding and application.

References

Aitchison, G. D. and Donald, I. B. (1956) Effective stresses in unsaturated soils. *Proceedings, 2nd Australia New Zealand Conference on Soil Mechanics,* 192–9.

ASAE (1989) Standard S313.2: Soil Cone Penetrometer.

ASAE (1990) Data Standard D497: Agricultural Machinery Management Data.

ASTM Standards (1982) Classification of soils for engineering purposes. D2488.

Bailey, A. C., Johnson, C. E. and Schafer, R. L. (1985) A compaction model for agricultural soil. *Proceedings of the International Conference on Soil Dynamics*, vol. 2. Auburn, AL, 203–12.

Bekker, M. G. (1956) *Theory of Land Locomotion.* University of Michigan Press, Ann Arbor.

Bernstein, R. (1913) Problems zur experimentellen motorpflugmechanik. *Der Motorwagen* 16.

Bishop, A. W. (1959) The principles of effective stress. *Teknisk Ukeblad* 39, 859–63.

Bishop, A. W. and Blight, G. E. (1963) Some aspects of effective stress in saturated and partly saturated soils. *Geotechnique* 13 (1), 177–97.

Bishop, A. W., Alpan, I., Blight, G. E. and Donald, I. B. (1960) Factors controlling the strength of partly saturated cohesive soils. *ASCE Res. Conference on the shear strength of cohesive soils.* Boulder, CO, 503–32 and 1027–42.

Black, D. K. and Lee, K. L. (1973) Saturating laboratory samples by back pressure. *Journal of Soil Mechanics and Foundations Division*, ASCE, 99 (1), 75–93.

Brackley, I. J. A. (1975) A model of unsaturated clay structure and its application to swell behaviour. *Proceedings 6th African Conference Soil Mechanics and Foundation Engineering*, Balkema, Rotterdam.

Brixius, W. W. (1987) Traction prediction equations for bias ply tyres. ASAE Paper No. 87-1622.

Burland, J.B. (1965) Some aspects of the mechanical behaviour of partly saturated soils. *Proceedings Conference on Moisture Equilibrium and Moisture Changes in Soil Beneath Covered Areas.* Butterworths, Guildford, 270–8.

Chancellor, W. J. and Korayem, A. Y. (1965) Mechanical energy balance for a volume element of soil during strain. *Transactions of the ASAE* 8, 426.

Cobb, D. E., Cohron, G. T. and Gentry, J. D. (1961) Scale model evaluation of earth moving tools. *1st International Conference of ISTVS, Proceedings*. Turin, Italy.

Croney, D., Coleman, J. D. and Black, W. P. M. (1958) Movement and distribution of water in relation to highway design and performance. *Highway Research Board, Special Report no. 40*. Washington DC.

Dwyer, M. J., Comely, D. R. and Evernden, D. W. (1975) Development of the N.I.A.E. handbook of agricultural tyre performance. *5th International Conference of ISTVS, Proceedings, vol. III*. Detroit-Houghton, Michigan.

Emori, R. I. and Schuring, D. (1965) Feasibility of model study in earthworking equipment, *ASAE Transactions* 298–300, 304.

Foster, C. R. and Ahlvin, R. G. (1954) Stresses and deflections induced by a uniform circular load. *Highway Research, Board Proceedings* 33, 467–70.

Freitag, D. R. (1965) A dimensional analysis of the performance of pneumatic tires on soft soils. *USAE Waterways Experiment Station Technical Report no. 3-688*.

Freitag, D. R. (1988) Principles of soil cutting and excavation: A review of Russian literature. *SAE Transactions* 97 (SAE Paper No. 880812).

Gill, W. R. and Vanden Berg, G.E. (1967) Soil dynamics in tillage and traction. US Department of Agriculture, Agricultural Research Service, *Agricultural Handbook no. 316*, p. 511.

Godwin R. J. and Spoor, G. (1977) Soil failure with narrow tines. *Journal of Agricultural Engineering Research* 22, 213–28.

Grisso, R. D. and Perumpral, J. V. (1985) Review of models for predicting performance of narrow tillage tool. *Transactions of the ASAE* 28(4), 1063–7.

Grisso, R. D., Perumpral, J. V. and Desai, C. S. (1980) A soil–tool interaction model for narrow tillage tools. *ASAE Paper 80-1518*.

Hettiaratchi, D. R. P. and Reece, A. R. (1967) Symmetrical three-dimension soil failure. *Journal of Terra Mechanics* 4(3), 45–67.

Hettiaratchi, D. R. P. and Reece, A. R. (1974) The calculation of passive soil resistance. *Geotechnique* 24(3), 289–310.

Hettiaratchi, D. R. P., Whitney, B. D. and Reece, A. R. (1966) The calculation of passive pressure in two-dimensional soil failure. *Journal of Agricultural Engineering Research* 11(2), 89–107.

Hodek, R. J. (1972) Mechanism for the compaction and response of kaolinite. *Joint Highway Research Project Report 36*. Purdue University, Lafayette, IN p. 269.

Holtz, R. D. and Kovacs, W. D. (1981) *An Introduction to Geotechnical Engineering*. Prentice-Hall, Englewood Cliffs, NJ.

Hvorslev, M. J. (1937) Uber die festigkeitseigenschaften gestorter bindiger boden (Physical properties of remolded cohesive soils). Translation No. 69-5, J. C. Van Tienhoven, US Army Waterways Experiment Station, Vicksburg, MS.

Hvorslev, M. J. (1960) Physical components of the shear strength of saturated clays. ASCE Research Conference on the Shear Strength of Cohesive Soils. Boulder, CO, 169–273.

Janosi, Z. (1961) An analysis of pneumatic tyre performance on deformable soils. *Proceedings 1st International Conference of ISTVS*. Turin, Italy.

Jennings, J. E. B. and Burland, J. B. (1962) Limitations to the use of effective stresses in partly saturated soils. *Geotechnique* 12(1), 125–44

John Deere (1980) *Factors In Earthmoving*. Moline, IL.

Karafiath, L. L. and Nowatzki, E. A. (1978) Soil mechanics for off-road vehicle engineering. *Trans. Tech.,* Clausthal, Germany; Gower Publishing Company, p.515.

Kawamura, N. (1985) Soil dynamics and its application to tillage machines. *Proceedings International Conference on Soil Dynamics.* Auburn, AL.

Knight, S. J. (1956) A summary of trafficability studies through 1955. *USAE Waterways Experiment Station*, TM No. 3–240, fourteenth supplement.

McKyes, E. (1985) *Soil Cutting and Tillage.* Elsevier, New York.

McKyes, E. and Ali, O. S. (1977) The cutting of soil by narrow blades. *Journal of Terramechanics* 14(2), 43–58.

NATO Reference Mobility Model (1979) *U.S. Army Tank Automotive Research and Development Command Technical Report no. 12503, VS02.* Warren, MI.

Nichols, Herbert L. Jr. (1976) *Moving The Earth: The Workbook of Excavation.* McGraw.

Nuttall, C. J. Jr. (1956) Scaled vehicle mobility factors – final report. *WNRE Report no. 18–2. USAE Waterways Experiment Station.* Vicksburg, MS.

Olson, R. E. (1963) Effective stress theory of soil compaction. *Journal of Soil Mechanics and Foundation Division, ASCE* 89(2), 47–98.

Osman, M. S. (1964) The mechanics of soil cutting blades. *Journal of Agricultural Engineering Research* 9(4).

Palmer, A. C., Kenny, J. P., Perera, M. R. and Reece, A. R. (1979) Design and operation of an underwater pipeline trenching plough. *Geotechnique* 29(3), 305–22.

Payne, P. C. J. (1956) The relationship between the mechanical properties of soil and the performance of simple cultivation implements. *Journal of Agricultural Engineering Research* 1.

Pavlics, F. (1966) Locomotion energy requirements for lunar surface vehicles. *SAE Paper no. 660149.*

Perumpral, J. V. and Desai, C. S. (1979) A generalized model for a soil–tillage tool interaction. *ASAE Paper no. 79–1546.*

Perumpral, J. V., Grisso, R. D. and Desai, C. S. (1983) A soil–tool model based on limit equilibrium analysis. *Transactions of the ASAE* 26(4), 991–5.

Plasse, R., Raghavan, G. S. U. and McKyes, E. (1985) Simulation of narrow blade performance in different soils. *Transactions of the ASAE* 28(4), 1007–12.

Proctor, R. R. (1947) Description of Field and Laboratory Methods. *Engineering News-Record.*

Reaves, C. A., Cooper, A. W. and Kummer, F. A. (1966) Similitude in performance studies of soil–chisel systems. *ASAE Paper No. 66–125.*

Reece, A. R. (1964) Theory and practice of off-the-road locomotion. *Institute of Mechanical Engineering Conference.* London.

Reece, A. R. (1965) The fundamental equation of earthmoving mechanics. *Institute of Mechanical Engineers Symposium on Earthmoving Machinery.* London, England, 8–14.

Roscoe, K. H., Schofield, A. N. and Wroth, C. P. (1958) On yielding of soils. *Geotechnique* 8, 22–53.

Rula, A. A. and Nuttall, C. J. Jr. (1971) An analysis of ground mobility models (ANAMOB). *Technical Report M-71-4.* USAE Waterways Experimental Station.

Schofield, A. N. and Roth, C. P. (1968) *Critical State Soil Mechanics. McGraw-Hill,*

London.

Söhne, W. (1953) Pressure distribution in the soil and soil deformation under tractor tyres. *Grundlagen der Landtechnik* 5, 49–63.

Söhne, W.(1956) Some basic considerations of soil mechanics as applied to agricultural engineering. *Grundlagen der Landtechnik* 7.

Söhne, W. (1958) Fundamentals of pressure distribution and soil compaction under tractor tyres. *Agricultural Engineering* May, 276–91.

Söhne, W. (1963) The mechanics of soil–vehicle systems with particular respect to agricultural tractors. *Grundlagen der Landtechnik* 17.

Sullivan, R. J. (1964) Earthmoving in miniature. *SAE Paper no. 897B.*

Terzaghi, K. (1943) *Theoretical Soil Mechanics.* John Wiley, New York.

Toll, D. G. (1990) A framework for unsaturated soil behaviour. *Geotechnique* 40(1), 31–44.

Turnage, G. W. (1972) Tyre selection and performance prediction for off-road wheeled-vehicle operations. *Proceedings of the Fourth International Conference of the International Society for the Terrain Vehicle Systems*, 1. Stockholm, Sweden.

Vanden Berg, G. E. (1966) Triaxial measurements of shear strain and compaction in unsaturated soil. *Transactions of the ASAE* 9, 460.

Vanden Berg, G. E. Cooper, A. A., Erickson, A. E. and Carleton, W. M. (1957) Soil pressure distribution under tractor and implement traffic. *Agricultural Engineering* 38, 854–5.

Vesic, A.S. (1973) Analysis of ultimate loads of shallow foundations. *Journal of Soil Mechanics and Foundations Division, ASCE* 99(1).

Voorhees, W. B. (1986) The effect of soil compaction on crop yield. *SAE Paper no. 860729.*

Waterways Experiment Station – Corps of Engineers (1954) Investigations of pressure and deflections for flexible pavements. *TM No. 3–323, Report 4*, Homogeneous Sand Test Section.

Wismer, R. D. and Luth, H. J. (1971) Rate effects in soil cutting. *SAE Paper no. 710179.*

Wismer, R. D. and Luth, H. J. (1972) Off-road traction prediction for wheeled vehicles. *ASAE Paper no. 72–619.*

Wong, J. Y. (1989) *Terramechanics and Off-road Vehicles.* Elsevier, New York.

Yong, R. N. and Fattah, E. A. (1976) Prediction of wheel–soil interaction and performance using the finite element method. *Journal of Terramechanics* 13(4), 227–40.

Yong, R. N. and Hanna, A. W. (1977) Finite element analysis of plane soil cutting. *Journal of Terramechanics* 14(3), 103–25.

Yong, R. N., Fattah, E. A. and Boonsinsuk, P. (1978) Analysis and prediction of tyre–soil interaction and performance using finite elements. *Journal of Terramechanics* 15(1), 43–63.

Zelenin, A. N. (1950) *Basic Physics of the Theory of Soil Cutting.* Moscow.

Zelenin, A. N., Balovnev, V. I. and Kerov, I. (1975) Machines for moving the earth. *Mashinostroenie*, Moscow, p.555. (USDA, TT79–52007, 1985.)

Zoz, F. M. (1970) Predicting tractor field performance. *ASAE Paper no. 70–118.*

Zoz, F. M. (1978) Predicting tractor field performance (updated). *ASAE Paper no. 87–1623.*

Chapter 3

Tractor Ride Dynamics

H. Göhlich

Introduction

The historical development of the tractor is characterized by the continuous growth of size and power and by the continuing enlargement of its application spectrum. Today, the tractor is not only used for conventional pulling-tasks, but has been developed into a universally applicable machine, becoming one of the most complex of all agricultural machines. The demands for increased performance and capacity resulted in higher driving speeds, wherever the performance of the task allowed it. Driving-dynamics problems arose at the same time. The main problem for conventional tractors is the vibrational behaviour caused by the accepted design, which has no suspension system. The only suspending elements for the entire vehicle are the low-pressure tyres, which have not been developed as a suspension element, but to create high pulling power and, today, low ground pressure.

The optimization of the dynamic behaviour follows two main aims. These are measures for the increase of driving comfort and those for driving safety. Comfort is not only a reference to general well-being, but is aimed at the fulfilment of working and medical demands. Vibrational stresses are involved, but so also are other disturbing stresses – for example, those caused by noise. The vibrational load still represents a real danger for the health of the driver, who is subjected to such non-lasting symptoms as increasing tiredness and decreasing efficiency, as well as lasting and later-developing damage (Mitschke, 1984).

A first approach to the decrease of vibrational loads was the improvement of the seat suspension, especially through the use of air-suspended seats. Air-suspended seats, however, could only solve the vibrational problem to a limited extent. The main disadvantage of any suspended seat is the fact that relative movements occur between the seat and the controls that still cause discomfort for the driver. Furthermore, most of the seats do not offer protection against horizontal vibrations.

Driving safety is mainly influenced by strong dynamic oscillations of the axle-load. Especially while carrying out high-speed transportation tasks on the road, or while driving with heavy rear implements, the front wheels are intermittently unloaded causing a temporary loss of ground contact. Loss of driving control may result.

General aspects of driving dynamics

Vibrational movements of tractors

While driving on a road, vibrational movements occur on the vehicle, mainly because of the unevenness of the road. Additionally, the 'non-ideal' form of the tyres (eccentricity, non-circularity) and the unbalanced masses of the wheels create self-excited vibrations.

Those vibrations are transferred through the vehicle into the driver, who can be submitted to a strong load. Furthermore, dynamic stresses in the constructional components of the vehicle are created. Simultaneously, the wheel loads can differ considerably. The loads on single wheels may be temporarily nil; this is especially dangerous when the steering front wheels are concerned, as the steering ability of the vehicle is temporarily lost.

The known vibrational modes are vertical, longitudinal and lateral vibrations as well as the rotational modes of pitch, roll and yaw (Fig. 3.1). While driving straight ahead vertical vibrations, and rotational movements around the lateral axis (pitch) and around the longitudinal axis (roll) are excited. Lateral vibrations as well as the longitudinal modes can be neglected for vibration technical analysis. The same is true for the yaw (rotational movements around the vertical axis). Yaw and lateral vibrations are caused in straight-ahead movement by lateral forces that are typical for vehicles with independent wheel suspension. With these vehicles (motor cars) the spring deflection of the wheel causes a shifting of the track-width and a changing of the king-pin angle, so that the tyres are subjected to lateral loads (Mitschke, 1984). Tractors, which are equipped normally with rigid axles, are not subject to this effect.

Among the most important characteristics of the vibrational behaviour of tractors are the natural frequencies of the vertical vibrations, the pitch and the roll. The value of these frequencies, as well as the amplification of the excitation at the resonance points, is determined mainly by the weight and the moments of inertia, and the properties of resilience of the tyres, because in general other suspension elements do not exist. The natural

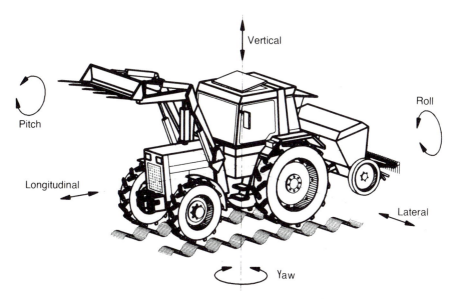

Fig. 3.1. Degrees of freedom for a tractor.

frequencies for tractors lie between 2.0 and 3.5 Hz. The maxima of the frequency response functions are in general high, because the only suspension elements – the tyres – possess a very low damping coefficient.

The excitation of vibrations

Self-excited vibrations

While driven, the vehicle is subjected to a vibrational excitation that can be divided into periodical and non-periodical parts. The unevenness of the road is in general a non-periodic function. Furthermore, it is stochastically distributed. Besides these excitations, every vehicle itself generates vibrational movements. This self-excitation occurs periodically and is a result of the rotational movements of the wheels and tyres. The main reason for this excitation is not the unbalanced mass, but the non-circularity of the wheels. The fundamental wave of the vibrational excitation can be described for every wheel with the following equations:

$$u = \hat{u} \sin(\omega t + \varphi_0) \qquad [3.1]$$

where

$$\omega = \omega_{wheel} = \frac{v}{r_{wheel}} \qquad [3.2]$$

(v is the forward velocity, r_{wheel} the radius of the wheel), and

$$u = \hat{u} \sin \left(\frac{v}{r_{wheel}} t + \varphi_0 \right) \qquad [3.3]$$

For the nth harmonic wave the natural frequency becomes:

$$\omega_n = n \omega_{wheel} \qquad [3.4]$$

The angle ψ_0 determines the rotation of the phase of the excitation (the position of the wheel) at the time $t = t_0$. The terms t_0 and φ_0 are called zero-phase time and zero-phase angle respectively (DIN 1311).

For tractors, the first harmonic and the second harmonic waves are of great influence. They can be interpreted as radial run-outs of first and second degree, as eccentricity and elliptical differences from the ideal roundness of the tyre. Non-circularities of higher degrees do exist, but they are of less importance, primarily because of their small amplitude, and secondarily because of their excitation frequency, which is much higher than the natural frequency of the vehicle. The influence of the lugs, for example, is a high-frequency excitation. The lugs do cause a periodic excitation, but this frequency is so high that the vibration behaviour of the vehicle is not influenced. The periodic excitation of the lugs, though, does cause a significant noise at the frequency of the lug impact.

If an excitation frequency f_n (or $\omega_n = 2\pi f_n$) is identical with a natural frequency of the vehicle, then a critical driving condition will commence, even with small excitation amplitudes, owing to the resonance vibrations. Since the excitation frequency is a function of the driving velocity, critical velocities do result. For a vertical normal frequency of $f_{vert} = 2.5\,Hz$ (typical for tractors) and a diameter of the front wheel of approximately 1.3 m (Beckmann, 1989), the critical velocity is commonly found to be between 35 and 40 km/h.

For tractors, which are getting faster and faster these days, self-excitation in this region of practical speeds yields in many cases a considerable decrease in driving comfort and safety (Kising, 1988).

Excitation by the road

Besides (periodic) self-excitation (mostly stochastical), excitation by the road is of significant influence on driving comfort.

As explained above, periodic signals can be divided into fundamental and harmonic waves, where the fundamental wave has the period T. This period T can be associated with the frequency $f = 1/T$ or the circular frequency $\omega = 2\pi f$. The identification of the harmonics of a periodic vibration is derived with harmonic analysis, which is formulated for the nth harmonic of a periodic signal as follows:

$$\underline{X}_n = \hat{X}_n e^{j\varphi_{0n}} = \frac{2}{T} \int_0^T x(t)\, e^{-j2\pi n f_1 t}\, dt \qquad [3.5]$$

The term $\underline{X}_n$ is a complex function (or a 'complex amplitude'). The angle between the real and the imaginary axis represents the phase position of the signal. For the characterization of the signal, the amplitudes (zero-amplitude) and the phase shifts (zero-phase angle) are examined over the frequency (amplitude and phase spectra).

Stochastic signals – that is, the excitation introduced by the road – can be regarded as a special case of periodic vibrations if they are based on a theoretical period length of infinity. From this assumption all of the relations for non-periodic vibrations are deduced. For such an infinite period length the frequency f (and the circular frequency ω) is converging to zero. Consequently, the higher harmonic waves are situated infinitely close together, and a continuous spectrum is created. The transformation into the spectral domain is carried out with the help of a Fourier Transform. The formula for the transformation for any signal $x(t)$ is:

$$\underline{X}(\omega) = \int_{-\infty}^{\infty} x(t)\, e^{-j\omega t}\, dt \qquad \omega = -\infty \ldots \infty \qquad [3.6]$$

The Fourier-transformed function is a symmetrical function; that is, it is mirrored on the axis $f = \omega = 0$. Between the signal in the time- and the frequency-domain a relation can be formulated with the help of the root-mean-square value (RMS):

$$T x_{eff}^2 = \int_{-\infty}^{\infty} x(t)^2\, dt = \int_{-\infty}^{\infty} \underline{X}(\omega)^2\, d\omega \qquad [3.7]$$

The complex Fourier-transformed $\underline{X}(\omega)$ yields again an amplitude and a phase spectrum, where the amplitude spectrum is normally indicated as the (one-sided) power spectral density (PSD) Φ:

$$\Phi(\omega) = |\underline{X}(\omega)|^2 + |\underline{X}(-\omega)|^2 = 2|\underline{X}(\omega)|^2 \qquad \omega = 0 \ldots \infty \qquad [3.8]$$

As opposed to the time signals $x(t)$, the unevenness of the road is not a function of time, but of displacement. The excitation by the road, u, is a function of the displacement s; the time t is involved through the driving velocity v. If v is constant it follows that:

$$u(vt) = u(s) \qquad [3.9]$$

that is,

$$vt = s \qquad [3.10]$$

Substituting s for t makes it possible to transfer the relations of the time-dependent values to the displacement-dependent values. For this reason the circular displacement frequency Ω is introduced as a value analogous to the circular frequency ω:

$$\Omega = 2\pi/L \qquad [3.11]$$

where L is the wavelength of the unevenness, and

$$\Phi(\Omega) = V\Phi(\omega) \qquad [3.12]$$

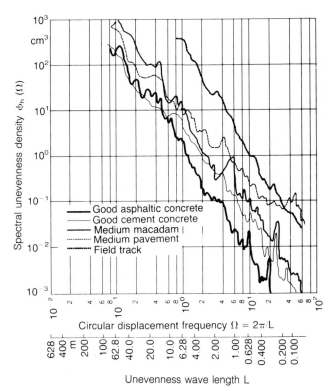

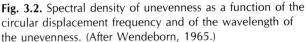

Fig. 3.2. Spectral density of unevenness as a function of the circular displacement frequency and of the wavelength of the unevenness. (After Wendeborn, 1965.)

Figure 3.2 shows typical spectra of road unevenness. Plotted on double logarithmic scales they represent a good approximation of a straight line (Wendeborn, 1965). For the approximation of the spectra of the unevenness of the displacement, the following formula is valid:

$$\Phi(\Omega) = \Phi(\Omega_0) \left[\frac{\Omega}{\Omega_0}\right]^{-w} \qquad [3.13]$$

Here Ω_0 is the reference circular displacement frequency (e.g. 1 unevenness m^{-1}). $\Phi(\Omega_0)$ can be regarded as the quality of the unevenness of the road.

The exponent w determines if the road has mainly higher frequency or lower frequency components. For a good field track, Braun (1969) gives values of $\Phi(\Omega_0) = 32\,\text{cm}^3$ and w = 2.25.

The properties of the distribution of uneveness wavelengths of roads can be based in almost all cases on a Gaussian normal distribution. According to Jungerberg (1984), this assumption is only valid if a minimum length of the road (0.5 km) is exceeded.

Influence of the wheelbase

A simple model of a vehicle is the single-track model shown in Fig. 3.3, in which the tyres are represented as simple spring–damper combinations. The excitation in the front and in the rear is identical, but between the excitation of the front and the rear wheels lies a delay time t_d that depends on the wheelbase l_{wb} and the driving velocity:

$$t_d = l_{wb}/v \qquad [3.14]$$

and

$$u_r(t) = u_f(t - t_d) \qquad [3.15]$$

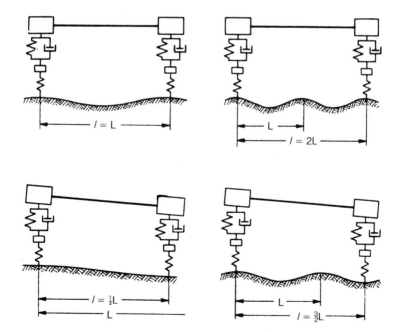

Fig. 3.3. Deduction of the critical velocity.

This is in contrast to self-excited vibrations, where no direct relation between the excitation of the different wheels can be found.

The model in Fig. 3.3 possesses two degrees of freedom: pitch and vertical movement. Consequently, two acceleration spectra result with two important resonances − one per degree of freedom. The Figure shows two special cases of the vibrational excitation. For a waveform-like road with an unevenness wavelength of $L = l_{wb}$ the excitation is a pure vertical one, while for $L = 2l_{wb}$ it becomes a pure pitch one. These two cases can be generalized: pure vertical excitation results for unevenness wave lengths of:

$$L = l_{wb}/n \qquad\qquad [3.16]$$

and pure pitch excitation for:

$$L = 2l_{wb}/(2n - 1) \qquad (n = 1, 2, 3, \ldots) \qquad [3.17]$$

To these unevenness wavelengths belong the corresponding frequencies $f = v/L$. For $L = l_{wb}$ the velocity v can be just so great that the resulting excitation frequency f equals the vertical frequency f_{vert} of the vehicle. Resonance thus results. The same is valid by analogy for the pitch, if $L = 2l_{wb}$ and if f equals the pitch frequency f_{pitch}.

Although, in general, wavelike roads do not exist, these relations are of importance for normal stochastic roads, because these spectra are continuous and therefore contain also the wavelength that has the properties referred to above. Consequently, critical driving states always exist if the following conditions are satisfied:

$$v = f_{vert}l_{wb}/n = l_{wb}/(nt_d) \qquad\qquad [3.18]$$

and

$$v = f_{pitch}2l_{wb}/(2n - 1) = 2l_{wb}/[(2n - 1)t_d]. \qquad [3.19]$$

These velocities are called critical velocities. Since the highest amplitudes of the unevenness spectra appear at the lowest circular displacement frequencies (or the longest wavelengths), n can be set to unity. The other cases with $n \neq 1$ are of minor importance. Often $v = f_{vert}l_{wb}$ is considered 'the' critical driving velocity. For a tractor with a wheelbase of $l_{wb} = 2.2\,m$ and a natural frequency of $2.5\,Hz$ this critical velocity is found at $20\,km/h$. The critical velocity that is related to the pitch has in this case a value of about $32\,km/h$ (for, typically, $f_{pitch} = 2\,Hz$).

The influence of the tyre on the ride dynamics

The dynamic behaviour of non-suspended vehicles on public roads is a function of many factors. The geometry of the vehicle, the quality of the tyres and the wheel rim, and the spring and damping behaviour of the tyres are of major importance for driving safety and comfort.

For the vibrational characterization of tyres, a distinction between the quasi-static and the dynamic condition has to be made. The following considerations refer to the spring and damping properties, as well as to the transfer and the amplification of the distortion influences.

Low-frequency vibration behaviour

Non-uniformities, unbalanced masses and the heterogeneity of the material create vertical dynamic forces on the rolling tyre, which are repeated with every revolution. The vehicle vibrations that are caused by these harmonic forces depend on the angular velocity of the wheel, which is a function of the driving velocity. If the excitation frequency is equal to the natural frequency of the vehicle, then considerable vertical amplitudes can be created in the resonance region. In this case one talks about a critical velocity. The critical velocity depends on the dynamic tyre-radius r_{dyn} and the degree of the harmonic excitation z_{exc}. With f_0 as the first natural frequency of the vibration system vehicle-mass–tyre and s as the wheel slip, the critical velocities result in:

$$v_{crit} = \omega_0 \frac{r_{dyn}}{z_{exc}} (1 - s) \qquad [3.20]$$

The harmonics of the fourth and higher degree are not important for driving safety. Since:

$$\omega_0 = \sqrt{\frac{c}{m}} \qquad [3.21]$$

the critical velocity depends on the system mass m and on the spring constant c (see the section on dynamic stiffness and damping below), that is on the inflation pressure. The presentation of the self-excitation follows by relating the maximum dynamic wheel load $F_{z,dyn,max}$ to the static wheel load $F_{z,stat}$. The result is called the dynamic load factor n:

$$n = 1 + \frac{F_{z,dyn,max}}{F_{z,stat}} \qquad [3.22]$$

If $n = 2$ the tyre loses ground contact; the region above $n = 1.6$ is considered critical. Braking and steering forces cannot be transferred satisfactorily. Fig. 3.4 shows this behaviour in the case of a rear tyre (18.4 × 38) with a radial run-out (maximum variation in radius) of ± 2 mm.

If the tyres are filled with water, which is common in agriculture, low fill-factors (below 75%) can create sudden uncontrollable vertical vibrations.

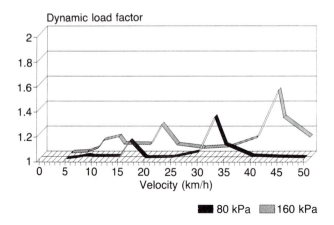

Fig. 3.4. Dynamic load factor as a function of velocity and inflation pressure ($m = 1550\,\mathrm{kg}$, tyre size 13.6×38).

For this reason, the velocity should not exceed 25 km/h. Fill-factors of 95% have a relatively stabilizing effect.

Vehicle vibrations in the lateral direction can influence driving comfort and safety considerably. Two cases have to be taken into account. On the one hand, low stiffness values in the lateral direction cause large vibration amplitudes for straight-ahead driving (Pacejka, 1988). They are excited by the unevenness of the road, heterogeneity of the tyres and lateral deviations of the wheel rim. On the other hand, the steering adjustment frequencies of the driver in turning can cause resonances in the lateral direction.

High-frequency vibration behaviour

Because of impulses caused by the lug contact and lug separation of deeply lugged agricultural tyres, high-frequency excitations are introduced into the chassis of the vehicle. The resulting vibrations create additional stress in the structure of the tractor and reduce the driving comfort. The excitation frequencies at the axle are of great importance. They are influenced mainly by the profile and the transfer-function of the tyre. Furthermore, the characteristic of the tyre in relation to resonance vibration − that is, its inherent damping − is relevant.

The lug impulse frequency f_{lug} is a function of:

● number of lugs (distance between the lugs);
● rolling radius;
● velocity.

It can be calculated from:

$$f_{\text{lug}} = \frac{vj}{7.2\pi\, r_{\text{dyn}}(1 - s)} \quad [\text{Hz}] \qquad [3.23]$$

where v is the true driving velocity in km/h;

j is the number of lugs;

r_{dyn} is the dynamic rolling radius in m; and

s is the slip (in the range 0 to 1).

In practice, the measured vertical resonance frequencies differ only slightly from one type of tyre to another (resonance range I, 16 to 20 Hz; resonance range II, 24 to 27 Hz; resonance range III, 30 to 40 Hz) (Fig. 3.5). The width of the resonance range, as well as the amplitudes, differ strongly, however. A higher load-index (representing the carcase stiffness) usually leads to bigger amplitudes. By contrast, a reduction of the inflation pressure and consequently a reduction of the tyre stiffness does not necessarily lead to a decrease of the excitation amplitudes. Fig. 3.6 shows the high-frequency behaviour of the same tyre as in Fig. 3.5, but with a lower inflation pressure. This decrease in pressure results in an amplification of the peaks of resonance by 70%; because of the load on the tyre, the outer parts of the lugs touch the ground and the excitation forces are directly introduced into the sidewall of the tyre (Siefkes and Göhlich, 1990).

Low water fill-factors have a high damping effect and reduce the peaks of the high-frequency excitation intensity. No differences can be found between an air-filled tyre and a tyre that is filled to the maximum with water ($V_{\text{water}}{:}V_0 = 0.95$).

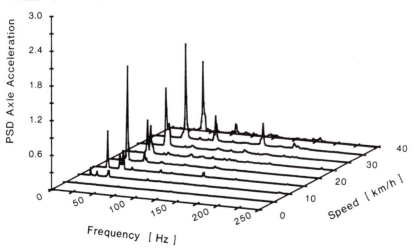

Fig. 3.5. Power spectral density of the lug excitation for a tyre inflation pressure of $p = 160\,\text{kPa}$ ($m = 2000\,\text{kg}$, tyre size $= 30.5 \times 32$).

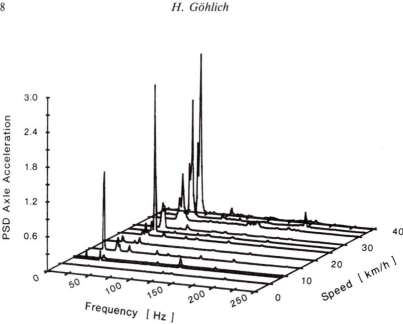

Fig. 3.6. Power spectral density of the lug excitation for a tyre inflation pressure of $p = 80\,\text{kPa}$ ($m = 200\,\text{kg}$, tyre size = 30.5 × 32).

Wear is in general an amplifying factor for high-frequency vertical vibrations. With continuous wear the vibration excitation of the tyre–vehicle system increases and driving comfort decreases in consequence. The research to date indicates that for a complete investigation of the properties and of the construction of the tyre, as well as the entire vibration behaviour of the vehicle, not only the new tyre but also the used worn tyre must be investigated. Measurements on agricultural tyres show that the intensity of the excitation in the vertical direction will increase strongly even with minimal wear (IAE Berlin, unpublished).

High-frequency excitation in the lateral direction is typical for tyres with an agricultural profile. The large lugs and the alternate directions of the lug contact create a tilting moment around the longitudinal axis and therefore lateral forces in the axle.

Dynamic stiffness and damping

As already shown for low-frequency vibrational behaviour, stiffness plays a major role for a tyre under load. It influences the frequency with which the vehicle vibrates; it also determines the maximum possible load. The damping influences the maximum possible amplitudes of the tyre–chassis-mass vibrational system.

There is a significant difference between static and dynamic stiffness as there is between static and dynamic damping. The static parameters are measured with simple force–displacement experiments, where the axle is fixed. By contrast, the dynamic parameters are determined with a pre-loaded, freely vibrating tyre.

For these considerations the non-linear continuum vibration model tyre is replaced by a single-mass vibration model with a spring and damping element. The solution of the differential vibration for a free, damped single-mass vibration model yields the natural circular frequency:

$$\omega^* = \omega_0\sqrt{1 - D^2} \qquad [3.24]$$

For very small damping $D \ll 1$ it is valid that:

$$\omega^* = \omega_0 = \sqrt{\frac{c_{dyn}}{m}} \qquad [3.25]$$

where c_{dyn} is dynamic stiffness in N/m; and
      m is mass in kg.

The stiffness is therefore:

$$c_{dyn} = \omega_0^2 m = 4\pi^2 f_0^2 m \qquad [\text{N/m}] \qquad [3.26]$$

Furthermore, the solution of the differential equation of the free damped one-mass vibration model yields the decay-rate:

$$\delta = \frac{r}{2m} \qquad [\text{Hz}] \qquad [3.27]$$

where r is the damping constant in Ns/m.

Multiplying this equation with the reciprocal of the natural frequency defines the dimensionless damping ratio:

$$D = \frac{\delta}{\omega_0} = \frac{r}{2m\omega_0} \qquad [3.28]$$

The vertical stiffness and the damping of the tyre are to a great extent non-linear. They depend on the physical values of load, inflation pressure, velocity and temperature. They also change with the excitation frequency, caused by the roughness of the route. Furthermore, differences in stiffness and damping can be determined that are due to different ways of tyre construction.

The stiffness of agricultural tyres has the following typical characteristics (Kising, 1988; Kising and Göhlich, 1988):

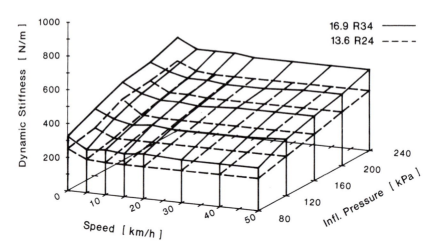

Fig. 3.7. Dynamic stiffness of a front and rear tyre as a function of the velocity and of the inflation pressure (front: *m* = 958 kg, tyre size 13.6 × 24; rear: *m* = 1560 kg, tyre size 16.9 × 34).

- between 0 and 10 km/h the stiffness decreases by approximately 20–25% (Fig. 3.7);
- above 10 km/h the stiffness is constant;
- increasing the inflation pressure will result in an increase of the stiffness;
- for higher workloads the stiffness will increase;
- cross-ply tyres possess a 10–20% higher stiffness than radial tyres of a comparable dimension;
- with increasing excitation frequency the tyre stiffens (Clark, 1982).

Typical characteristics can also be identified in the damping behaviour of large-volume agricultural tyres (Kising, 1988; Kising and Göhlich, 1988):

- often the damping increases slightly betwen 0 and 5 km/h by about 5% (Fig. 3.8);
- above 5 km/h damping declines asymptotically; for 50 km/h the limit at 30% of the original value is reached;
- damping declines with increasing inflation pressure;
- higher workloads lead to an increase of the damping;
- cross-ply tyres show higher damping at low inflation pressures than comparable radial tyes; at high inflation pressures no differences can be found.

The lateral stiffness and damping of a tyre are also highly non-linear. Four typical characteristics can be identified:

- the lateral stiffness is about 35–50% of the vertical stiffness;

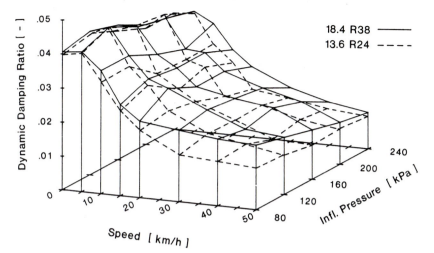

Fig. 3.8. Dynamic damping constant of a front and rear tyre as a function of the velocity and of the inflation pressure (front: $m = 970\,kg$, tyre size 13.6×24; rear: $m = 1620\,kg$, tyre size 18.4×38).

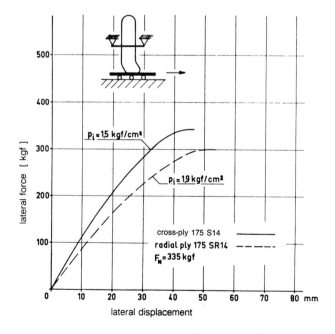

Fig. 3.9. Behaviour of the lateral stiffness of a radial and a cross-ply tyre. (After IAE Berlin, unpublished.)

- Stiffness declines with increasing lateral load (Fig. 3.9);
- cross-ply tyres possess a 10–20% higher stiffness than comparable radial tyres;
- the damping in the lateral direction is about 90–100% of the damping in the vertical direction (Clark, 1982).

Closely related to lateral stiffness is the torsional stiffness around the vertical axis. Fig. 3.10 shows the torsional moment versus the twisting angle.

The natural frequencies of the tyre vibration system are related to the stiffness. In the vertical direction, the first natural frequencies lie between 1.7 and 3 Hz (4.5 Hz for twin tyres); in the lateral direction, between 1 and 2.2 Hz (up to 3.2 Hz for twin tyres).

The evaluation of vibrations

Driving safety and loads

An evaluation of the suspension displacements, the forces, the accelerations and the like is needed in order to make a statement about the driving behaviour

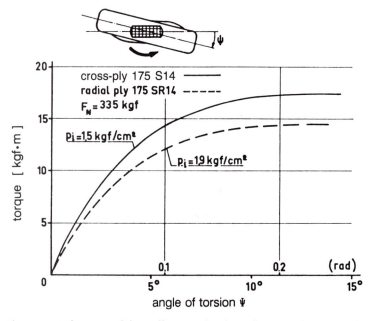

Fig. 3.10. Behaviour of the stiffness under the influence of a torsional moment about the vertical axis, radial versus cross-ply tyres. (After IAE Berlin, unpublished.)

of a vehicle. Since it is not possible to make a statement about steering quality based on the dynamic wheel-load alone, a criterion for driving safety is the quotient of the root mean value of the dynamic wheel-load to the static load. This quotient is called the wheel-load factor $n*$.

$$n* = \frac{F_{z.dyn}}{F_{z.stat}} \quad \text{with} \quad F_{z.dyn} = F_{total} - F_{stat} \qquad n* \geq 0 \qquad [3.29]$$

With increasing wheel load factor $n*$ the probability of losing the ground contact increases. According to Ulrich (1983), $n*$ should not be bigger than 0.33. For a normal distributed oscillation of the wheel-load, positive ground contact for $n* = 0.33$ exists for 99.73% of all amplitudes. Particularly for tractors with rear implements, this value is often exceeded on the front wheels, since the front axle is statically unloaded.

For the specification of structural components of vehicles, like wheels or wheel-bearings, the amount of shocks within the vehicle vibration is particularly important. For this reason the dynamic load factor has been introduced into design calculations (see the section above on low-frequency vibration behaviour).

Often the maximum dynamic force is limited to:

$$F_{z.dyn.max} = 3F_{z.dyn.eff} \qquad [3.30]$$

This means that, with the assumption of a normal distribution, only in 0.15% of all cases do forces greater than $F_{z.dyn.max}$ appear. For tractors, the dynamic load factor may exceed 2.

Driving comfort

The driver is subjected, through the seat, the feet and the hands, to translational vibrational movements. Additionally, angular accelerations, which are due to the pitch and the roll, affect the driver. These loads may threaten the driver's health, especially over long periods of exposure. For agricultural vehicles, the vertical vibration accelerations are considerably reduced by suspended seats.

The different evaluation of vibrations of different frequencies through the human being results from the construction of the different organs of the human body, which represent vibration systems themselves, possessing natural frequencies and resonances. Many of these natural frequencies lie in the region between 4 and 8 Hz (Göhlich, 1987), and are therefore sensed as especially uncomfortable.

For the evaluation of vibration that influences the human body, three main criteria have to be differentiated:

1. reduced comfort boundary;

2. fatigue decreased proficiency boundary;
3. exposure limit.

This differentiation can be found in ISO Directive 2631 (1978). In this directive the frequency-dependence of the human body is evaluated with respect to the vertical and the horizontal directions of input. The human body reacts very sensitively in the vertical (backbone) direction z to frequencies between 4 and 8 Hz and in the horizontal directions x and y to frequencies below 2 Hz. Endurance decreases with increasing duration of exposure. These physiological facts are reflected in the evaluation curves based on measurements with pilots and drivers. For the vertical direction the evaluation curves are proportional to the acceleration in the high- and in the low-frequency region. Below 4 Hz and above 8 Hz the intensity is evaluated to be less.

This evaluation of vibration can also be done with the help of the similar VDI Directive 2057 (1987). In this directive weighting functions for the different vibrations are introduced. These are based on experiments with human beings, who were submitted to sinusoidal excitations of constant acceleration amplitude with different frequencies. The subjects were asked to evaluate the oscillations in a subjective manner. Weighting curves of equally evaluated vibrational strength did result from these measurements. Weighting functions B for the acceleration spectra can be deduced. For example, the function for the vertical seat acceleration is:

$$
\begin{array}{ll}
B_{\text{seat}} = 4.48 & \text{for 0 to 0.2 Hz} \\
B_{\text{seat}} = 4(2\pi f)^{1/2} & \text{for 0.2 to 4 Hz} \\
B_{\text{seat}} = 20 & \text{for 4 to 8 Hz} \\
B_{\text{seat}} = 1000/(2\pi f) & \text{for } > 8 \text{ Hz}
\end{array}
$$

Multiplying the RMS spectra by this function B will give the weighted spectra. From this result a new, weighted RMS value can be deduced – the so-called k-value. To obtain a complete evaluation of all the vibrational influences, all of the k-values have to be linked by the following equation:

$$
K = \sqrt{\sum_{1}^{n} k_i^2} \tag{3.31}
$$

For up to 10 minutes a K-value of 2 is supportable, but above this the well-being is influenced. For higher values the vibration will be strongly felt. In the same way well-being can be disturbed if low K-values are allowed to be imposed for a longer time.

K-values, though, cannot be used to evaluate shock-bearing vibrations. Shocks can be introduced if the suspension travel is not long enough and the 'end stop' is reached. They are sensed as especially uncomfortable. To characterize them the crest-factor F_c is used:

$$F_c = \frac{z_{max}}{z_{RMS}} \qquad [3.32]$$

where z_{max} is the maximum shock acceleration; and

z_{RMS} is the RMS value of the vibration acceleration.

Vibration models for tractors

An example of a vehicle model was shown in Fig. 3.1, and represents the effective degrees-of-freedom of the vehicle and its different component systems, using known notations.

Figure 3.11 shows a simplification of the real vehicle that considers only those parts that are relevant to its dynamic behaviour; that is:

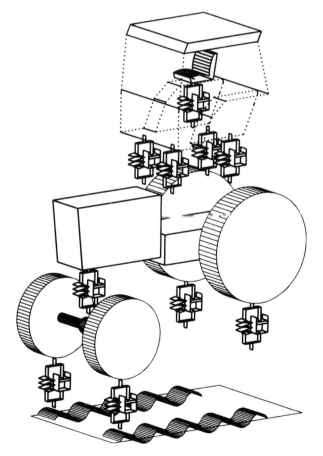

Fig. 3.11. Model of a tractor.

- masses and moments of inertia;
- stiffness and damping constants; and
- the geometrical dimensions (that is, wheelbase and the coordinates of the centre of gravity).

A further reduction of the model to a plane single-track model has to be considered as unsatisfactory. This is especially true for a judgement of the real motional behaviour of the vehicle and its accelerations as well as the driver's stresses.

For a plane single-track model, two statistically identical excitations (for example, two phase-shifted road tracks) would result in statistically identical vibrational behaviours of the system. The deficiency of the single-track model becomes obvious as soon as the two extreme cases, pure pitch and pure roll, are considered. A three-dimensional model leads to a system which allows us to consider the resulting contribution of the roll and the movements of the suspended axle, as well as the evaluation of those vibrations in respect to the physiological effects on the human being.

Another decomposition of the vertical/pitch substituting system is shown in Fig. 3.12. Here the rigid mass of the vehicle has been separated according to the front and rear axle loads. The conditions valid for the identity between the real vehicle and the substitution model according to Fig. 3.12, are:

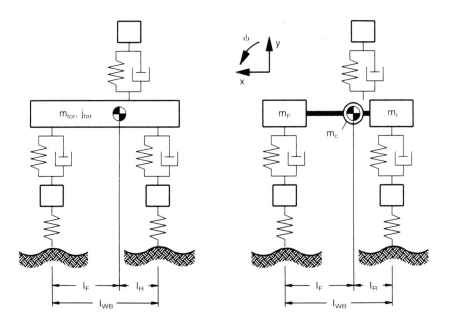

Fig. 3.12. A simplified plane tractor model.

$m_f + m_r + m_c = m_{tot}$ (constant rigid mass)
$m_f l_f = m_r l_r$ (constant coordinates of the centre of gravity)
$J_{tot} = m_f l_f^2 + m_r l_r^2$ (constant pitch moment of inertia)

Generally these conditions can only be fulfilled by the introduction of the so-called 'coupling mass'. However, its value often is of little significance – for example, for numerical simulation, because it neither aids understanding (it even might be negative) nor does it simplify the differential equations of motion. For the evaluation and optimization of the vibrational behaviour of agricultural tractors the vertical and the pitch vibrations are of basic importance. However, the other degrees of freedom have to be considered in order to explain and describe the generation of these movements completely (cf. Fig. 3.11). This is especially so when detailed optimization processes are required to maximize the driver's comfort.

Techniques for vibration isolation

Besides the 'classic' remedy for the isolation of vibrations – the suspended seat – other solutions have to be considered. There are three basic systems, each possessing specific advantages and disadvantages:

- suspended cabin;
- front axle suspension;
- implements as dynamic vibration absorbers.

Cabin suspension uses the principle of the vibrational isolation of the driver's workplace (Kauß, 1981). One important advantage of this solution is the much higher total suspended mass compared to the seat–driver system, leading to considerably lower natural frequencies. In such a case, a simplification of the seat's suspension system becomes possible. A further improvement of driver comfort results from the avoidance of relative movement between the driver's body and the controls. Further progress in the utilization of electric and hydraulic transmission elements on tractors means that connection of the operating controls to the unsuspended parts of the tractor will become less difficult.

A stable suspension construction is needed to ensure kinematically exact guidance of the cabin as well as to fulfill the requirements of the law concerning overturning and resistance to continued rolling of the vehicle (Council Directive, 79/622/EEC). A suspended front axle will lead to an improvement in driving stability and safety because of the decrease in the dynamic front axle forces (see Fig. 3.13; Weigelt, 1987). The vibration stresses on the driver depend to a considerable extent on the position of the seat with reference to the axles. A mid-mounted cabin normally represents an optimum position with respect to the suspended front axle. Furthermore, such a cabin

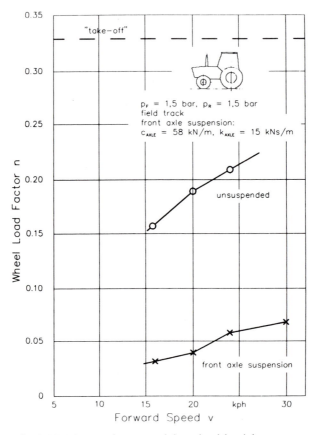

Fig. 3.13. Measured curves of the wheel load factor.
(After Weigelt, 1987.)

position will result in an increase of the front axle load, which will consequently stabilize the front part of the vehicle (Fig. 3.14).

The third solution is represented by an absorber system using mounted implements. The absorption principle presupposes a simple spring−mass system, which will only be effective for one specific operating frequency. Under these conditions the system will be completely calm. Because of the varying implement masses and the vehicle excitation frequency range the system will be broadband and will need additional damping (see Fig. 3.15).

Much more practicable is an active system, allowing a complete absorption by automatic control of the electro-hydraulic lift. A digital controller reduces the dynamic front axle loads, leading to considerable increase in driving safety and driving comfort. It has been proved that the use of control systems of this kind will permit doubling the road speed while maintaining the same vibrational loads (Bergmann, 1987). The use of mounted masses for

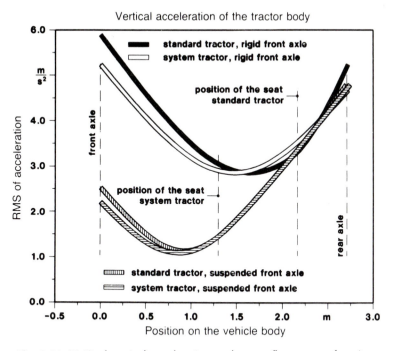

Fig. 3.14. RMS of vertical acceleration at the seat flange, as a function of the seat position.

absorption has a decisive significance for the exclusion of critical driving conditions during the transport of mounted implements. However, such a device does not solve the vibration problems of empty tractors or tractors towing trailers.

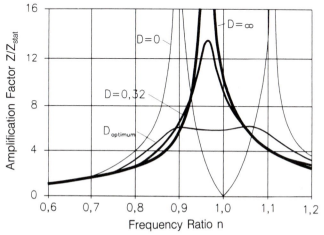

Fig. 3.15. Transfer function of an absorber for different dampings.

Acknowledgements

The author expresses his appreciation to Peter Pickel, Tjark Siefkes and Christian Kaplick, scientific staff members of the Institute of Agricultural Engineering, Technical University of Berlin, for their contribution to this study.

References

Beckmann, D. (1989) Normung landwirtschaftlicher Reifen. Lecture to the 7th VDI/MEG Colloquium (Reifen landwirtschaftlicher Fahrzeuge), 27–28 April, Munich.

Bergmann, E. (1987) Entwicklung von Hard- und Softwarewerkzeugen zur Leistungs- und Einsatzoptimierung von Traktoren. *Fortschritt-Berichte* 14 (39). VDI-Verlag, Dusseldorf.

Braun, H. (1969) 'Untersuchung von Fahrbahnunebenheiten und Anwendung der Ergebnisse.' Dissertation, Technical University of Berlin.

Clark, S. K. (1982) *Mechanics of Pneumatic Tires*. US Department of Transportation, Washington DC.

Council Directive 79/622/EEC. On the approximation of the laws of the Member States relating to the roll-over protection structures of wheeled agricultural or forestry tractors (static testing).

DIN 1311. *Schwingungslehre* (German Standard no. 1311 on mechanical vibrations).

Göhlich, H. (1987) *Mensch und Maschine. Lehrbuch der Agrartechnik,* vol. 5. Parey-Verlag, Hamburg/Berlin.

IAE Berlin (unpublished) Unpublished results from the Institute of Agricultural Engineering of the Technical University of Berlin.

ISO 2631 (1978). Leitfaden für die Bewertung von Schwingungsein-wirkungen auf den menschlichen Körper als ganzes. (A guide for the evaluation of human exposure to wholebody vibration.)

Jungerberg, H. (1984) 'Ein Beitrag zur experimentellen und numerischen Simulation von Traktorschwingungen.' Dissertation, Technical University of Berlin.

Kauß, W. (1981) 'Aktive, hydraulische Schwingungsisolierung des Fahrerplatzes ungefederter, geländegängiger Fahrzeuge.' Dissertation, Technical University of Berlin.

Kising, A. (1988) *Dynamische Eigenschaften von Traktor-Reifen*. VDI Verlag, 14(40). Dusseldorf.

Kising, A. and Göhlich, H. (1988) Ackerschlepper-Reifendynamik (Parts 1–3). *Grundlagen der Landtechnik* 38 (3–5), 78–87, 101–36, 137–68.

Mitschke, M. (1984) *Dynamik der Kraftfahrzeuge,* vol. B: *Schwingungen;* 2nd edn. Springer Verlag, Heidelberg/Berlin/New York/Tokyo.

Pacejka, H. B. (1988) *Modelling of the Pneumatic Tyre and its Impacts on Vehicle Dynamic Behaviour.* Carl-Cranz-Gesellschaft, Oberpfaffenhofen.

Siefkes, T. and Göhlich, H. (1990) Dynamic characteristics of very large agricultural tyres. *Papers in Agricultural Engineering '90.* Berlin.

Ulrich, A. (1983) 'Untersuchungen zur Fahrdynamik von Traktoren mit und ohne Anbaugeräte.' Dissertation, Technical University of Berlin.

VDI Directive 2057 (1987) Beurteilung der Einwirkung mechanischer Schwingungen auf den Menschen. (Assessment of the effect of mechanical vibrations on human beings.)

Weigelt, H. (1987) Schwingungseigenschaften vorderachsgefederter landwirtschaften Traktoren. *Fortschritt-Berichte* 14(33). VDI-Verlag, Dusseldorf.

Wendeborn, J.O. (1965) Die Unebenheiten landwirtschaftlicher Fahrbahnen als Schwingungserreger landwirtschaftlicher Fahrzeuge. *Grundlagen der Landtechnik* 15(2), 33ff.

Chapter 4

Stability of Agricultural Machinery on Slopes

A.G.M. Hunter

Introduction

Agricultural machinery is required to operate over terrain which can be very steep. Unlike roads, agricultural land can have sharply varying contours and a rough uneven surface. Stability problems arise with tractors, trailers, and all types of farm machine which travel over the land, and few farms are free from sloping areas that are dangerous. Safe operating limits are easily exceeded. Unfortunately, the driver has little or no information on these limits, and overturning accidents occur as a result.

During the 1950s there was concern in many countries about the increasing number of deaths caused by overturning tractors, and this started a search for safety improvements. The safety cab was adopted as a primary measure because it offered protection during a roll-over. Legislation was introduced into the UK in 1967, and into many other countries at a similar time, requiring new tractors to be fitted with safety cabs (MAFF and DAFS, 1969).

Fitting safety cabs did not, however, address the causes of overturning, and research into machine stability also continued. It was clear that the subject required study in greater depth than manufacturers were able to offer, and even now there are almost no data available on safe operating limits. Legislation, on the other hand, was introduced into the UK in 1974 to make safety at work a legal obligation for all concerned, from the manufacturer right through to the farm employer and his employee. This certainly includes safety from overturning tractors, and a full understanding of stability on slopes is therefore needed.

The purpose of this chapter is to review the research which has been carried out, to indicate its present state of development, and to suggest the direction of future needs. First, the various aspects of machinery behaviour on slopes are covered, leading to calculation methods and results for the static stability of different types of machine. The question of stability under dynamic conditions is then tackled, followed by methods and approaches to the problem of assessing safe operating limits for machines on slopes.

Overturning accidents

Numbers and costs of accidents

The seriousness of overturning accidents is most easily measured by the numbers of fatalities occuring, though the fatality rate can be expressed in several ways. Manby (1963), for example, found a comparable fatality rate for driving tractors to that for driving road vehicles, of approximately one fatality per 9,000 tractors or road vehicles per year. Jensen (1980) calculated the number of fatal tractor accidents in Wisconsin during the 1970s as approximately 15 per 100,000 farm workers per year. Perhaps more relevant, a number of studies illustrate that the fatality rate from overturning accidents is a high proportion of the fatality rate from all causes on farms (Table 4.1).

The table suggests that the number of fatalities caused by overturning has reduced over the years. The HSE (1986) study in Britain confirms this, showing a continuous fall from over 60 fatalities in 1966 to 20 in 1984 for drivers of self-propelled machines. The fall is attributed to the marked

Table 4.1. Numbers of fatalities on farms.

Fatalities caused by overturning, per year	Fatalities, all causes, per year	Period of study	Country of study	Source
38	nr	1958–1962	England and Wales	Manby (1963)
35	114	1967–1969	England and Wales	Marples (1971)
25	nr	1969–1971	England and Wales	Chisholm (1972)
10	74	1981–1984	England, Wales and Scotland	HSE (1986)
200	nr	1965 ⎫	W. Germany	Schwanghart (1978)
53	nr	1977 ⎭		
17	49	1970–1974 ⎫	Nebraska, USA	Schneider (1980)
9	39	1975–1979 ⎭		

All figures are quoted or calculated from the sources given, and rounded to whole numbers.
nr – not reported.

reduction in deaths due to overturning tractors, corresponding with the introduction of safety cabs. Schwanghart (1978) plots the fatality rate per 10,000 tractors per year against the year of tractor manufacture (rather than the year of fatality), and offers clear evidence of a sharp fall in the fatality rate after cabs were introduced in West Germany.

However, non-fatal accidents are also serious and are far more numerous. Baligand (1978) reports 40 fatalities with tractors during one year in France, but 350 serious injuries, and 1,662 accidents which resulted in at least one day off work; the average length of time off work was 42.8 days. Monk *et al.* (1986) estimated the numbers of farm accidents in Great Britain 15 years after safety cabs were first introduced. They made a study of every reported accident in 1981/82, and carried out a questionnaire survey of unreported accidents on 850 farms; they also estimated the resulting costs of these accidents to the British economy. Their figures for tractor overturning accidents alone are 10 fatalities costing a total of £576,000 and 1,234 non-fatal accidents costing £3.3 million (at 1982 prices).

Types of overturning accident

Willsey and Liljedahl (1969) looked specifically at non-fatal overturning accidents, before considering how to reduce their numbers. Of 145 cases, they found that 120 were side overturns, 21 were rear overturns, and four were front overturns; travel over an obstacle, and working with a front-loader, were principal causes. Chisholm (1972) noted that implements or trailers were attached to tractors in about three-quarters of overturning accidents, and tractor speed exceeded 13 km/h before overturning in about one-third of cases, implying high levels of kinetic energy to be dissipated. A quarter of the tractors with cabs rolled more than one complete revolution after overturning. Half of the tractors were out of control before overturning, mainly sliding down a hillside (Whitaker, 1973).

Owen and Hunter (1983) classified a total of 560 reported accidents in the UK, considering primarily the distinction between a limitation of the tractor and a misjudgement by the driver. For example, overturning on a steep slope was taken as exceeding the limitations of the tractor, while driving into a ditch was taken as misjudgement by the driver. They found that tractor limitations were exceeded in 55% of cases, covering steep slope, high speed, rough ground, and loss of control; they also confirmed the high percentage of cases with equipment attached to the tractor (Table 4.2).

A number of case studies were then taken in detail, including computer simulations of the circumstances leading to overturning, in order to examine the critical factors (Hunter and Owen, 1983). It was found that a tractor with a stability limit of 33° had overturned sideways on a slope of only 5° as a result of excess weight on the front loader, and that another tractor with

A.G.M. Hunter

Table 4.2. Classification of tractor overturning accidents.

Cause of accident	Equipment				Total	
	Trailed	Mounted	None	Not recorded	Number	%
Stability loss						
(i) Slope exceeds tip angle	36	55	2	2	95	17[a]
(ii) Speed high	24	22	10	–	56	10[a]
(iii) Ground rough	12	18	4	–	34	6[a]
Control loss	69	42	12	2	125	22[a]
Driver's misjudgement	50	64	29	2	145	26
Miscellaneous	38	39	28	–	105[b]	19
Total	229	240	85	6	560	
%	41	43	15	1		

[a]Tractor limitations exceeded in 55% of total accidents.
[b]Miscellaneous accidents included traffic accidents (28) and driverless tractors (37).

mounted fertilizer spreader had overturned on a slope between 3° and 6° under the combined influences of speed and rough ground. A silage trailer, with a stability limit of 29° when empty, had a stability limit of 21° when full which resulted in overturning on uneven land at the edge of the field. A tractor fitted with a mounted rotary mower overturned on a slope of 12° when cornering on a radius of 3.5 m at 13 km/h, the centrifugal effect reducing the stability limit from 31°. A tractor overturned rearwards when working with drag harrows on a slope of 16° as a result of 0.6 g acceleration after momentary loss of grip and then a recovery. Finally, a tractor and trailed spreader overturned on a slope exceeding 25° when the combination reached the limits of both stability and control, resulting in a multiple roll (Fig.4.1).

Data from all of these studies have been used to build up a general picture of the relationship between slope and stability (Table 4.3). Slope angle measured in degrees above horizontal is used throughout this chapter as the

Table 4.3. Slope and stability limits.

Slope				
Degrees	Percentage	Gradient	Description	Unstable machine
0	0		Flat	Tractor, cornering at high speed
5	9	1 in 11	Gentle	Tractor, cornering at full lock, e.g. with mounted mower
10	18	1 in 6	Medium	Tractor, with big round bale on front loader (tipping sideways) Articulated-steer loader (some configurations)
15	27	1 in 4	Steep	Tractor, with heavy mounted sprayer (rearing) Silage trailer, full or unevenly loaded
20	36	1 in 3	Very steep	Trailed equipment, e.g. lime spreader, dung spreader, or slurry tanker (but less stable when part-full)
25	47	1 in 2	Excessive	Tractor, with standard wheel track
30	58		Extreme	Tractor, with wide wheel track Transporter, and other specialist mountain machinery

Fig. 4.1. Extensive damage to a tractor after a multiple roll accident.

principal unit, but percentage slope and gradient have also been included in the table for reference; the latter two units are simply alternative ways of expressing the tangent of the slope angle. The slope descriptions are intended as qualitative expressions of slope as it affects machine stability and, in the last column, typical examples of stability problems are given.

Performance and safety

Steep slope performance

Grecenko (1984a) defined the problem of operating machinery on steep slopes as one of performance as well as safety from overturning. Performance may be limited by functional shortcomings within the machine, by inability to cover the terrain at an adequate speed, or by damage caused to the crop or the ground. Schwanghart (1978) refers to estimated slope limits for different crops as follows: sugar beet 7°, potatoes 11°, cereals 14°, forage 17°, and grazing 24°. These limits depend on the machinery used and are mainly related to traction limits for pulling the harvesting equipment. Dettwiler and Gammenthaler (1975) propose a set of slope classifications for land, while recognizing that mechanization methods and equipment may change over the years (Table 4.4).

Table 4.4. Slope and mechanization.

Slope (degrees)	Ease of mechanization
0–6	No mechanization problems
7–10	Limit for root crop harvesting; slight problems with cereal and forage harvesting
11–14	Increasing problems with cereal and forage harvesting; slight problems with haymaking
15–19	Limit for cereal harvesting; forage harvesting very difficult; increasing problems with haymaking
20–27	Limit for haymaking

It is clear that grass is the only practical crop on steeper land, and Ott (1978) goes on to show the effect of increased slope in reducing efficiency. He considers the total work hours to harvest one unit of forage, from mowing right through to transport: based on a four-wheel-drive tractor, the hours increase by 45% as the slope range increases from 0°/10° to 14°/19°. Better traction and excellent stability are available using specialized mountain transporters, but unfortunately these can impose a substantial time penalty:

the 45% increase extends to 73% or even 445% if the machine is small and hand loading must be used. On the other hand, the transporter can be used directly uphill and downhill on slopes up to 31° under the best ground conditions and then, where the topography is suitable, the time penalty can remain small.

Grass imposes a variable surface for tyre adhesion, depending on grass maturity and the recent weather. Grecenko (1984b) calculates a variation between 35° and 16° in the limiting slope for a four-wheel-drive tractor climbing uphill, based on field results. He also reports that the steer angle required to maintain travel along the contour line may exceed 6°, and the drift angle of the whole machine may exceed 10°, on side slopes between 14° and 30°; as a result, the uphill wheels may ride on the uncut crop and the downhill wheels may drift onto the cut swath. If heavy forage harvesting equipment is trailed behind a two-wheel-drive tractor, Hunter (1981) finds that it may start to slide bodily downhill on a slope of only 6°. When the tractor is on its own, it will start to slide out of control on a steeper slope; depending on the ground conditions, it will then slew round and overturn sideways, or slew right round and continue to slide backwards downhill (Crolla and Spencer, 1984).

Quigley (1986) has developed an empirical model for tyre braking force coefficient on grass, based on extensive experimental work. Variations in grass conditions, and tyre and vehicle parameters, are found to cause variations in braking force coefficient between 0.93 and 0.23, implying a very wide range of slopes on which a machine can slide out of control. He also finds that braking force coefficient drops markedly when sliding starts, leading to a rapid build-up of speed.

Safety developments

There have been serious attempts to adapt or develop designs specifically for slopes. A minimum stability limit of 40° has been achieved for the transporter and the four-wheel-drive mountain tractor, notwithstanding the penalty of the small payload. This coincides with Molna's (1963) recommendations for Norway though, ironically, the stability limit of the typical farm tractor has reduced from 40° to 33° in the intervening years (Hunter, 1982a).

Ribetou (1984) has tackled the problem of payload by adapting existing machinery to produce a forage harvesting system for slopes, based on the four-wheel-drive tractor. Dual wheels are used to improve overturning stability as well as wheel grip on grass, and the tread direction on the front wheels is reversed for added wheel grip coming downhill (Zwaenepoel *et al.* 1987). The tractor is fitted with a cutter-bar mower which is front mounted for visibility and is light enough not to affect stability. A trailed forage pickup wagon is fitted with a powered axle for climbing slopes up to 17°, and it has wheeled outriggers which augment the static stability of the full machine from 28° to 47° (Hunter, 1985a). Similar developments in mechanization are taking place in Switzerland (Ott, 1989).

Owen (1987) points out that machinery manufacturers could provide easier means of widening the wheel track for improved stability. He also suggests that a trailed machine could roll right over, without overturning the tractor, through careful combination of swivel hitch on the drawbar and correct shape for the trailer body. This is in contrast to adjusting axle heights on side slopes as a means of keeping the tractor level (Wooley, 1921; Meyer, 1956), although this is more concerned with efficient traction than with stability. Patented devices are claimed to prevent overturning, for example by cutting the fuel to the engine (Dillman, 1973), or by shifting the centre of gravity of the machine (Biller and Johnson, 1987). Schwanghart (1971), however, concludes that cut-out devices and others such as self-extending props are not normally practical on tractors.

Aids to safety

The universal safety aid to provide operator protection is the safety frame or cab. Methods of testing its strength were developed in Sweden and evaluated in the UK using full-scale overturning trials (Manby, 1963). Tractor cab testing is now an integral part of the OECD tests for agricultural tractors and includes both impact tests using a swinging pendulum, and crushing tests using hydraulic rams: in both cases the energy to be absorbed during the test is based on the tractor mass (OECD, 1988). The cab can only pass the tests if a prescribed zone of clearance is maintained around the tractor seat, if no cracks appear in the cab strucure, and if the cab can withstand the crushing force.

Warning instruments in the cab include a simple slope indicator which is commonly fitted to mountain tractors, and can easily be fitted to any type of tractor. The use of warning devices is advocated by Murphy and Johnson (1984) in order to aid the driver, but the need to relate the information to a mathematical model of vehicle stability is also recognized (Murphy *et al.*, 1985). The problem is handled by Gibson *et al.* (1981) with a stability alarm, which is a mechanical analogue of the articulated tractor on which it is fitted; this design takes steering into account. Wray *et al.* (1984) have developed a stability alarm for an articulated loader shovel, and tested it under working conditions. In their case the instrument is an electronic analogue which is able to update its stability model, including the effects of bucket loads and steer angle, from sensors; a further refinement is to add derivative feedback for predicting dynamic effects. Spencer *et al.* (1983) introduced a meter for measuring the critical slope for control loss. This allows the driver to assess ground adhesion during a sliding test before travelling onto the steeper parts of the land, and has proved effective in farm trials (Owen and Hunter, 1987). An alarm to signal the need for engaging four-wheel drive on steeper land was also developed by Owen (1986).

Finally, education and training remain an important aid to improving safety. Considerable information on maintaining stability is contained in Clark (1974) published by one of the principal tractor manufacturers. In the UK, the Health and Safety Executive publish advisory leaflets and two educational films giving details of overturning accident causes and preventive measures (HSE, 1986). A teaching aid for demonstrating a full range of accident types with scale models, and for explaining some of the advanced concepts involved in the analysis of stability, has been developed by Hunter (1982b).

Stability limits

Scope of stability analysis

Early work on stability concentrated on rearwards overturning and is the basis of the subject in modern teaching texts (such as Liljedahl *et al.*, 1979).

Table 4.5. Variables affecting the stability of tractors.

Type	Variable
Static	Centre of mass (c.g.) height Track width Wheelbase
Dynamic	Inertia Tyre damping ratio Tyre spring rate
Initial conditions	Initial velocity of c.g. Angular velocity Acceleration of c.g. Angular acceleration
Driver controlled	External forces Steering Braking Engine torque
Terrain properties	Gross coefficient of traction Lateral tyre force coefficient Tyre slip angle Rolling resistance Topography Slope

By 1983, Grace *et al.* found in their review that rearwards overturning due to incorrect hitching, acceleration, slope, and obstacles, had been widely covered; sideways overturning while cornering had been analysed; and several researchers had developed models for overturning on a side slope under dynamic conditions. The number of studies on articulated tractor stability, their particular interest, was limited, and they included few references to static stability analysis.

In 1987, Kim and Rehkugler reviewed the scope of modelling techniques used by different researchers, commenting that the specific computer models which enabled rapid progress to be made from the 1960s onwards are now being superseded by general dynamics models which are available commercially, such as DADS and ADAMS. They give a brief summary of variables to incorporate into a stability model (Table 4.5). They consider that the rigid body parameters of the tractor can be treated relatively easily, but tyre deformations, tyre/ground interactions, and ground deformations are the elements

Table 4.6. Types of stability model.

Type	Model	Sources
Static	Prediction of zero wheel load Static conditions Rigid wheels Plane slopes	Blankenship *et al.* (1984) Hunter (1985b) Reichmann (1972a, b) Spencer (1978, 1988)
Quasi-static	Prediction of zero wheel load Steady state conditions Rigid wheels Plane slopes	Coombes (1968) Daskalov (1971 *et seq.*) Grecenko (1983) McKibben (1927) Newland (1981)
Dynamic	Prediction of overturning Dynamic conditions Rigid/flexible wheels Surface roughness / obstacles on slopes	Daskalov (1971 *et seq.*) Feng and Rehkugler (1986) Grecenko (1983) Hunter *et al.* (1988) Larson and Liljedahl (1971) Pershing and Yoerger (1969) Spencer and Gilfillan (1976) Van Deusen (1967) Zakharyan (1972)
Overturning	Dynamics of overturning Dynamic conditions Deformable wheels and tractor structure Deformable ground on banks/slopes	Chisholm (1981) Davis and Rehkugler (1974a, b) Rehkugler (1978) Schwanghart (1978)

of a model which are most difficult to treat, or have the least reliable data.

In this chapter, a number of different stability models will be presented but, rather than concentrate on the modelling technique, the aim will be to discuss the contents of the models. The summary in Table 4.6 is given in order of progression from models incorporating the most simplified elements to the most general, and the references are grouped as closely as possible by type.

The polar diagram

Reichmann (1972a, b) was prompted to analyse the static stability of transporters in Austria because existing methods could not predict the stability limits of these machines on steep slopes. It was thought that a frequent cause of overturning was exceeding the static limit of stability when loading a stationary vehicle. He looked at three-wheel vehicles initially, since these were statically determinate when one of the wheel loads fell to zero, and found that the stability limit for all orientations of the vehicle on a plane slope could be plotted as a three-sided stability boundary on a polar (r, ϑ) diagram. Stability limit was plotted as slope percentage represented by radius (r); heading direction was plotted as polar angle (ϑ); and the unstable wheel was identified as one side of the stability boundary. By this means he introduced a method for calculating and presenting stability for all orientations, including rearwards and sideways, all on the same diagram.

Another statically determinate case is the four-wheel vehicle when the loads at any two of the wheels have fallen to zero, and this produces a four-sided stability boundary. The typical agricultural machine, however, has two bodies with an oscillation pivot between them and in this case partial instability will occur with zero load at only one of the wheels. The general model for such a machine is shown in Fig. 4.2. The body-fixed axes 1,2,3 are oriented at a heading angle α to the uphill direction on a slope β. The wheels contact the ground at points a,b,c,d, the bodies have centres of mass s_1, s_2, the overall centre of mass is at s, and l is any point on the pivot axis. No assumptions are made about symmetry of the machine, or even about the direction of the pivot axis e. In order to reduce the equations to a determinate system, Reichmann assumes that all wheel forces A, B, C, D are parallel to the gravity forces G_1, G_2, and G.

The resulting set of plots on a polar diagram for a transporter is given in Fig. 4.3. The total stability boundary for zero load at two wheels, for example AB, has four corner points where load remains at one wheel only, for example a. The partial stability boundary for zero load at one wheel, for example A, has points of minimum stability, for example m_A at 44° (98%). The four superimposed boundaries 1, 2, 3, 4 show that a higher pivot on this machine tends to increase stability at the rear wheels but reduce

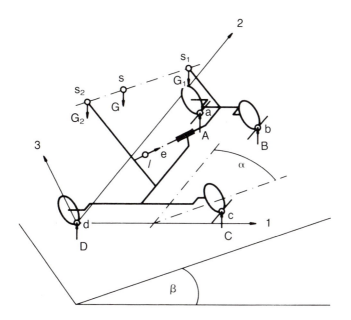

Fig. 4.2. General model of a four-wheel tractor comprising two bodies with an oscillation pivot.

stability at the front wheels. It is clear that there is very great stability for rearwards overturning, at a heading of 180°, and lower stability for sideways overturning, at a heading of 90° or 270°, but minimum stability on rear wheel D at a heading of 240°, or on front wheel A at a heading of about 320°. The slight asymmetry of the plots is explained by the transmission torque which crosses between the two bodies on this four-wheel-drive machine and is included in the model.

Spencer (1978), following the same approach, dealt with the problem of an indeterminate force system by assuming that wheel side forces, A_1, B_1, C_1, D_1, are directly proportional to wheel normal forces, A_3, B_3, C_3, D_3, and that, for a two-wheel-drive tractor, longitudinal forces A_2, B_2 on the front wheels will be zero while forces C_2, D_2 on the rear wheels will be equalized by the differential in the axle. He added into his model the effect of a trailer loading the hitch point, and the stability of the trailer itself, but he also included criteria for the total combination of two machines losing ground control, based on wheel side forces or longitudinal forces reaching the value for limiting friction (Fig. 4.4)

The mininum stability point for rear wheel D, m_D at 42° (90%) represents one component of wheel load, D_3, falling to zero. Under this condition, the two-wheel-drive tractor will roll freely downhill; all three components of wheel load will not necessarily reach zero until the hypothetical

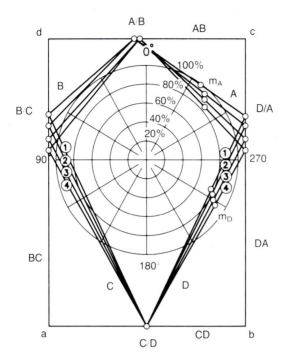

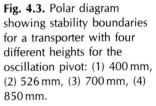

Fig. 4.3. Polar diagram showing stability boundaries for a transporter with four different heights for the oscillation pivot: (1) 400 mm, (2) 526 mm, (3) 700 mm, (4) 850 mm.

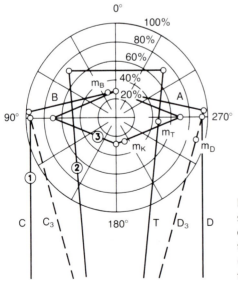

Fig. 4.4. Polar diagram showing stability boundaries for a two-wheel drive tractor with towed silage trailer: (1) tractor, (2) trailer, and (3) control boundary for tractor and trailer.

boundary D is reached. The minimum stability point for sideways overturning of the trailer, M_T at 25° (46%), is considerably less than for the tractor and thus the trailer will overturn first. However, the actual limitation is likely to be loss of control either downhill, with the minimum point m_K at 15° (26%), or uphill, with stability and control loss coinciding for front wheel B at the minimum point m_B, also at 15° (26%) in this example.

Reichmann's comprehensive account of this subject concentrates on the mathematical aspects, and also includes a large number of plots representing the effects of parameter changes in the design of several different vehicles. Spencer develops programs for stability and control of numerous vehicle types, later compiled into a menu-driven program by Gillham (1985). In the main, the programs are validated using scale models on slopes, but a full-scale radio-controlled tractor has also been used (Owen and Spencer, 1980).

Articulated steer machines

Gibson *et al.* (1971) developed a simple model of articulated forestry machines, considering only the rear wheels, and using geometrical methods to determine the partial stability limits. The model was able to take account of shifts in the tyre contact points under load, but treated the front axle mounting as a simple pin-joint, thus neglecting torques about axes perpendicular to the oscillation pivot. Marked reductions in stability with increasing steer angle were found, and the value of steer angle was shown to determine the heading angle at which minimum stability occurred. Results for a skidder were later compared with those for a forwarder carrying a range of payloads (Gibson and Biller, 1974).

A further model was developed by Blankenship *et al.* (1984) for a skidder with chassis oscillation; that is, with the oscillation pivot located at the articulation joint in the chassis, which they term double articulation. This model considered partial stability at the front as well as the rear wheels, and total sideways stability of the whole machine; the effect of load in the winch cable at the rear of the skidder was also added. Although methods of statics rather than geometry were used, the oscillation pivot was still treated as a pin-joint; predicted angles and cable forces were within 10%–20% of results from scale model trials.

Polar diagrams were used to plot the partial stability results from scale-model experiments by Kaloyanov *et al.* (1974), and both Reichmann (1972b) and Spencer (1988) extended their theoretical models to include articulated machines. Spencer found that, after solving for the partial stability boundaries of the front body, transformation to a new set of axes was needed to obtain solutions for the rear body when steered. This was successfully validated by Finlayson (1989) using a scale model with four different steer angles of 0°, 10°, 20°, and 40°. He found that all prediction errors for stability limit were less than 1.8° at all heading angles. He also used geometric methods to calculate the total stability boundaries (Finlayson, 1990).

Hunter (1988) adapted Spencer's model to include three alternative locations and alignments for the pivot axis: on the front chassis, on the rear chassis and on the front axle. He examined the worst stability cases for a digger/bucket loader. For example, with an unloaded bucket the minimum stability point m_B on one front wheel was 31° (60%) for two equivalent machines with a fixed chassis and with an articulated chassis, but was only 17° (31%) with the articulated machine steered to 40° (Fig. 4.5). With the bucket loaded and extended to full reach, the minimum point reduced to only 9° (16%) and shifted to rear wheel C. Locating the pivot for axle oscillation proved to be the least stable design in this study, but which alternative of front or rear chassis oscillation was the most stable depended on the loading conditions; a compromise design was therefore indicated. It was also found that one of the wheels could become so stable that it was never less stable than any of the other wheels, and thus the four-sided polar plot reduced to a three-sided one. This is easily visualized as a development of Fig. 4.5 in which the side for wheel A vanishes. Under these loading conditions the partial stability of one of the other wheels was normally very low.

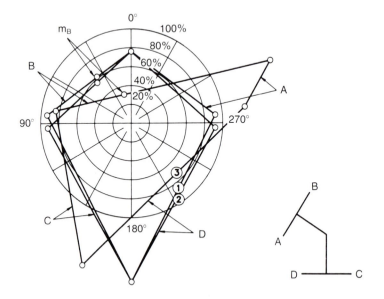

Fig. 4.5. Polar diagram showing the effect on stability of articulated steering: (1) fixed chassis vehicle, (2) pivoted chassis vehicle, (3) pivoted chassis vehicle at full steering articulation.

Fluid-filled tankers

In a sprayer tank mounted on a tractor, and especially in a trailed tanker, the mass of fluid is likely to dominate the stability of the machine. The centre of gravity position of the fluid is determined by the shape and attitude of the tank with respect to gravity: the centre of gravity coordinates are thus variable and are functions of both slope and heading angle. In addition, the mass of fluid in the tank depends on its initial fill and on the amount discharged during work. Hunter (1986) completed a centre of gravity analysis for circular and rectangular cross-section tanks, considering the summation of fluid elements for all tank attitudes in space, and for all levels of fluid fill.

He went on to examine the stability of trailed tankers with both shapes of tank, and with transversely mounted tanks (Hunter, 1985b). Since tanker fill was a variable which was additional to slope and heading angle, the polar plot was no longer adequate, and a contour plot was used to present the results. For the case shown in Fig. 4.6, sideways stability was greatest when the tank was empty at E, and over 10° less when the tank was full at F. Least

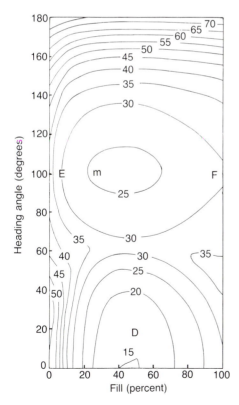

Fig. 4.6. Contour plot showing stability limits for a tanker with fill as a variable.

stability occurred at m because of the increased freedom for the centre of gravity to shift when the tank was only 30% full, and in spite of the reduced fluid mass. Additional work indicated the effects of the tanker on tractor control, a part-filled tanker putting minimal load on the drawbar at D, and severely limiting the slope which the tractor could climb (Hunter, 1984). The predictions were validated using scale-model and full-scale trials.

Dynamic stability

Acceleration effects

McKibben (1927) was concerned with rearwards overturning stability, deriving equations to predict the effects of acceleration after rapidly engaging the clutch, and defining the influence of centre of gravity position; he was also interested in the effect of hitch load and hitch position. His work included full-scale experiments on level ground, with the driving wheels bogged down, and on an uphill slope. Coombes (1968) used similar analysis, defining the critical slope as the one on which the tractor, or tractor with mounted implement, could maintain a constant uphill speed while the front wheel load tended to zero. He calculated the effect of towing a trailer uphill, showing that a high hitch point reduced the critical slope.

Radial acceleration while cornering at constant speed was also analysed by Coombes, with the lateral inertia force added to the equations for sideways instability. He derived solutions for both partial tipping at one rear wheel and total tipping at two side wheels, and demonstrated the sharp reduction in stability caused by increased speed and reduced turning radius. This was confirmed in full-scale trials carried out by Owen (1980a) using tractors mounted with a mower which was offset to the side, and with a hay tedder at the rear (Owen, 1981); additional trials (Owen, 1980b) on a tractor fitted with cross-ply tyres or radial tyres showed no effect due to tyre type compared with the effects of speed. Full overturning was prevented by an outrigger, but when a tractor wheel lifted it did so very sharply.

Newland (1981) studied an articulated machine and pointed out that cornering at speed on a slope did not represent steady-state conditions because of the changing direction of the radial acceleration with respect to the gravity vector, as heading angle changed. Therefore, momentary instability was not a sufficient condition for overturning. He calculated the effects of increasing steer angle, and increasing speed, on reducing the stability limit, and he considered the effects of design parameter changes. Kelly and Rehkugler (1980) examine the sharpness of the stability boundary for a tractor with maximum steering rate when turning uphill from the contour line, and demonstrate some conditions where wheel lift-off does not result in ultimate overturning.

Daskalov (1971, 1972, 1973, 1974, 1986) confines himself to general methods of analysis through which he contributes valuable insight into the study of dynamic stability. The dynamic equations of motion are regarded as the basis for stability analysis; static equilibrium is seen as a special case; and the effects of acceleration due to engine drive or steering are particular cases. Considering the tractor as a rigid body, he allows it to be acted on by external forces in addition to gravity, and employs the Euler equations of motion to define the dynamic motion, plotting the solutions on a polar diagram. The crucial concept which he proposes is that stable behaviour under dynamic conditions can be described as a locus of operating points inside the stability boundary. The margin of stability at any time can therefore be defined by the distance of the dynamic operating point from the static boundary, which is fixed by the initial geometry of the vehicle.

For example, when operating on a slope of 15° (30%) at a constant speed along a 55° heading angle, the operating point is p_0, and the locus of points along all heading angles is circle 1, well inside the static stability boundary (Fig. 4.7). If the tractor now accelerates in a straight line at acceleration f, then p_0 shifts to p_f and circle 2 is the new locus. The effects of adding the inertia force to the gravity force are to increase the effective slope by $(f/g \cos \beta)$, and to reduce the effective heading angle to 30°. The front wheels

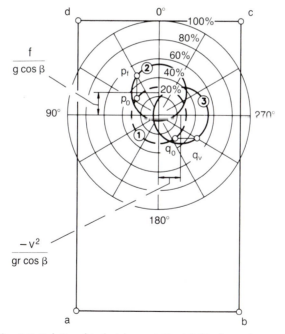

Fig. 4.7. Relationship between static stability boundary and locus of operating points on a 30% slope: (1) static conditions, (2) constant acceleration, (3) constant velocity cornering.

are less stable as a result. Similarly while cornering to the left at speed v on
a constant radius r, q_0 shifts to q_v on locus 3. The effective slope increases
by ($v^2/gr \cos \beta$), and the effective heading increases from 215° to 245°.
Wheels A and D are less stable as a result. Should the dynamic operating
point make an excursion right out to the boundary, then the machine has
reached the dynamic stability limit.

Overturning prediction

Schwanghart (1971) considered the kinematic roll motion of the tractor on
travelling over a bump of known height and then related angular velocity
to the speed of the tractor, assuming that the tractor wheel was rigid.
Grecenko (1983) chooses a standard flat-topped ramp and makes simple
calculations of maximum roll angle and maximum lift-off velocity at a rear
wheel. He includes stiffness and damping of the tyre, and finds reasonable
agreement with experiments. This leads to a graph of reducing safety margin
on slopes, with increasing speed, calculated on the basis of his level ground
data. Pershing and Yoerger (1969) developed a nine degree of freedom computer simulation of a tractor on a sideslope. The tractor was modelled as a
main body with front axle and four separate wheels, each with stiffness and
damping along three coordinate directions. Their standard obstacle was a
sinusoidal hump and they found that the rear wheel maintained ground
contact on all slopes up to 27° (50%) at a speed of 1.3 m/s. The rear wheel
lost ground contact on a slope of 14° (25%) at 2.0 m/s, and on level ground
at 2.7 m/s. An implement, mounted on the uphill side, improved stability.
    Larson and Liljedahl (1971) used piecewise linear spring rates, with
damping included, for calculating the normal loads on the front and rear
tyres in their model of a tricycle tractor. Tyre tractive force, lateral force,
rolling resistance, and slip were modelled from empirical data for tyres on
soil, and the standard obstacle was again a sinusoidal hump. A summary
of their results for five different slope angles, and three different obstacle
heights, is given in Table 4.7, showing the sharp reductions in both speed
and height allowable as slope increases. Their criterion for overturning was
that the kinetic energy of rotation about the tipping axis had equalled the
potential energy needed to reach the point of balance. This is shown as a
stability boundary on a phase plane type of plot; for example the results at
2.2 m/s over a bump 229 mm high on level ground fall below the boundary
1, while the results at 3.3 m/s are above and lead to overturning (Fig. 4.8).
It was found in full-scale validation trials that the tractor was considerably
more stable than the simulation predicted, probably due to inadequate
modelling of tyre forces and tyre–ground interaction.
    Feng and Rehkugler (1986) develop a full overturning model which
includes single wheel lift-off followed by lift-off at two side wheels. Their

Table 4.7. Simulated overturning situations.

Slope angle	Bump height (mm)	Speed (km/h)
0	75	8, 16, 24, 32
	150	4, 8, 12
	225	8, 12
7.5	75	8, 16, 24
	150	8, 12
	225	8, 12
15	75	8, 16, 20
	150	4, 8
	225	4, 8
22.5	75	4, 8, 12
	150	4, 8
	225	4, 8
30	75	4

The tractor overturned for the last test condition on each line above.

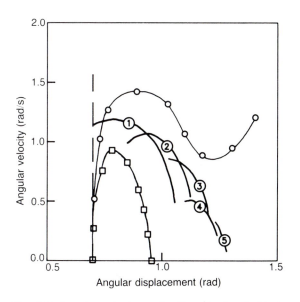

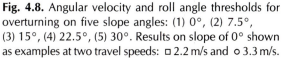

Fig. 4.8. Angular velocity and roll angle thresholds for overturning on five slope angles: (1) 0°, (2) 7.5°, (3) 15°, (4) 22.5°, (5) 30°. Results on slope of 0° shown as examples at two travel speeds: □ 2.2 m/s and ○ 3.3 m/s.

interest is in riding over an obstacle while steering a prescribed course on
a side slope, and they suggest that corrective steering or acceleration/
deceleration on the part of the driver can be used after wheel lift-off to prevent
ultimate overturning. Goldberg *et al.* (1989) have used a rollover simulator
to examine human response times for such evasive actions. Sestak *et al.* (1989)
are interested in overturn prediction based on the acceleration response of
a mountain mower on crossing a series of obstacles laid out in a line. Field
trials uphill, downhill, and along the contour line, on slopes of 17° and 30°,
are used to provide data for a computer model, with overturning energy used
as the stability criterion.

   Potential energy will be contributed by side slope, and kinetic energy by
the roll motion, but kinetic energy will show random variation due to the
machine dynamics on rough ground. Zakharyan (1972) handled this random
behaviour by taking an excursion of three standard deviations (3σ) from the
expected value (encompassing 99.7% of values in a normal distribution) as
the upper limit of energy available for overturning. Spencer and Gilfillan
(1976) developed the idea further saying that a safe excursion had to be limited
by first passage time, because even a single overturning represented failure.
A maximum value could therefore be calculated for angular velocity $\dot{\vartheta}_a$ of
the machine. This value would determine the available margin for potential

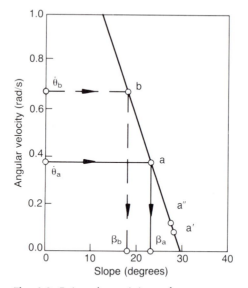

Fig. 4.9. Points determining safe
operating slope based on alternative
criteria for maximum allowable angular
velocity: (a) first passage time, (a') 3σ
excursion, (a'') 4σ excursion, (b) first
passage time on rougher ground.

energy, and hence the safe operating slope β_a (Fig. 4.9). Rougher ground and higher speed would increase the variance in angular velocity, and thus the maximum value for the same first passage would be higher $\dot{\vartheta}_b$, giving a lower operating slope β_b.

Van Deusen (1967) used a statistical technique for simulating vehicle motion over rough ground, when predicting behaviour of the vehicle on the moon. He included stiffness and damping at the wheels, at the suspension, and at the payload, and considered both non-yielding and yielding surfaces, although the effects of slope were not considered. Overturning forwards was more likely than sideways, so he estimated the probability of pitch angle exceeding the overturning value, assuming a normal distribution. The model was later extended to take account of wheel lift-off, a non-linear effect, and overturning was then predicted at a vehicle speed where the linear analysis predicted a 0.01 probability of overturning. Ground roughness was characterized by spectral density of profile height y in the form $S_y(n) = Kn^{-2}$ where K is a roughness constant and n is space frequency, a concept later used by Hunter (1979) to define a rapid method of measuring roughness data using an accelerometer.

It is interesting that wheel lift-off on its own was not sufficient for overturning in the simulation. In fact, the bouncing motion could develop to the extent that all four wheels lifted clear of the ground for 50% of the time before overturning occurred. Experiments carried out by Hunter *et al.* (1988) with an instrumented two-wheel trailer on sloping grass fields were designed to measure lift-off at the uphill wheel only. The durations of lift-off were used to calculate kinetic energy inputs contributing to roll motion. For each run, the additional energy needed to reach the total value for overturning would define the stability margin, which was then estimated as a percentage of the total energy and plotted against travel speed and slope (Fig. 4.10). At 2.5 m/s the stability margins were 74% on a slope of 18° and 98% on a slope of 13°; that is, on slopes of two-thirds and one-half the static tip slope β_s respectively.

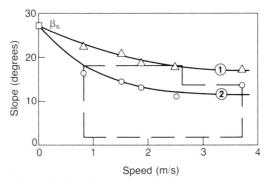

Fig. 4.10. Experimental results for slope as a function of speed at two levels of stability margin: (1) 74%, (2) 98%.

Overturning dynamics

Davis and Rehkugler (1974a, b) developed a general three-dimensional model
of the tractor, comprising the main body, front axle, rear wheels, clutch and
engine, which simulated all spatial motions. Tyre forces were derived from
radial spring rate, and traction and lateral force coefficients, taking account
of slip; for rough ground, an enveloping tyre model was used, but the ground
itself was treated as undeformable. Wheel lift-off was included, and external
forces could be added, from a draft implement for example. Validation results
were obtained with a one-twelfth scale model which was rolled onto a ramp
to initiate overturning. The model was later used by Rehkugler (1978) to
simulate a four-wheel-drive articulated-steer tractor overturning down a bank.
    Chisholm (1981) restricted his simulation to motion in two dimensions,
primarily to cover slipping and overturning sideways down a short steep
bank. He included elasto-plastic deformation of the rear wheels, the cab
and the ground, and he used a variable slip/slide model for transverse forces
at the ground, which depended on the current value for ground penetration
at the contact point. At each computation step he evaluated forces and
deflections for all contact points, and summed the effects of these to solve
for an incremental advance in the overturning motion. Full-scale validation
trials were carried out with a tractor which was set up for a range of parameter
combinations, 30 tests in all. The results show good agreement between
simulation and experiment for lateral, vertical and roll motion (Fig. 4.11).
    Schwanghart (1973) was concerned with the parameters of tractor
geometry which would inhibit continuous rolling down a slope, after an
overturn had started. His model also considers impact and deformation at
the wheel and at the cab, and includes ground deformation. His criteria for
non-continuous rolling are based on simulations using mean data from a range
of tractors, although the variability of soil conditions would also affect actual
behaviour.

Assessing stability

Measuring static stability

Measurement of centre of gravity position is compulsory under Code I for
the Official Testing of Agricultural Tractor Performance, although optional
under Code II (OECD, 1988). Its position can be used to estimate the total
stability limits, knowing the test figures for wheelbase and track widths. A
tilt table measurement of stability is not required and, outside Norway, does
not appear to be carried out anywhere (Hunter, 1982b). Tilt-table data are
very revealing, showing particularly the low values of stability for trailed

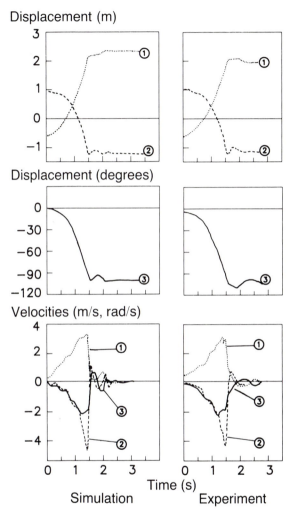

Fig. 4.11. Simulation and experimental results for one overturn showing time histories of rigid-body motion: (1) lateral, (2) vertical, (3) roll.

machinery compared with tractors, and the large reductions in stability caused by adding loads to empty machines (Table 4.8). The difference of 5° in stability at either side of the baler is also worth noting since the machine looked symmetrical from the outside. These are stability variations which can be predicted easily at the design stage using the wide range of computer models available. Computer models also make it practical to explore parameter values which would lead to optimum stability.

Table 4.8. Stability limits of some machines measured at the Scottish Centre of Agricultural Engineering during 1988–1989.

Machine	Features	Track width (mm)	Tip angle (degrees)
1. Tractors of different types			
Two-wheel drive	no driver	1630	36
Four-wheel drive	no driver	1840	42
Equal axle weights	no driver	1630	20
Compact	no driver	940	34
	with driver	940	30
Compact (bucket and	bucket down	940	30
digger fitted)	bucket up	940	26
Articulated	steer straight	1020	33
2. Trailed machinery			
Dung spreader	full load	1700	26
Round baler	empty, door fully open	1910	30/25[a]
Bale transporter	full load, transport position	1620	19
	full load, elevated position	1620	16
Silage trailer	full load, normal wheel track	1780	20
	full load, extended track	2150	24
3. Other machinery			
Bucket loader	full load, bucket down	1912	32
	full load, bucket up	1912	19
Digger/bucket loader	empty, bucket down	1910	33
	full load, bucket up	1910	22
Forestry forwarder	full load	1960	28

[a]Asymmetrical machine – tip values given for left and right sides.

An alternative measurement method using weighpads has been introduced by Spencer *et al.* (1985). The method is based on measuring weight transfer on a moderate slope, and then predicting the slope angle where an uphill wheel will carry no load (Fig. 4.12). When analysing the experimental data, it was decided that a sinusoidal model was best for predicting wheel load reduction with slope, corresponding to a two-dimensional approximation for the machine. Later work confirmed that this provided reasonable accuracy for stability prediction and there are now sufficient data to specify the measurement procedure, the weighpads, and the range of suitable machine types (Hunter *et al.*, 1990). The method is portable, offers the prospect of stability measurement by all machinery manufacturers at minimum cost compared with installing a tilt table, and is now to be standardized. A

Fig. 4.12. Stability prediction on a preset slope using weighpads to measure weight transfer.

microprocessor-based stability measurement unit which embodies this method has been developed by Owen and Hunter (1988).

Dynamic tests

Schwanghart (1978) proposed a test of performance which would determine a stability margin based on energy levels measured during the bounce when crossing a standard obstacle. Grecenko (1983) introduced a similar test which was later incorporated into a comprehensive series of tests for a tractor. In addition to the hazard of sideways overturning, hazards which occurred when cornering, accelerating, slipping, skidding, and sliding downhill, are to be tested (Fig. 4.13; Grecenko, 1986). The effects of mounted and trailed implements are included, and both dry and wet ground conditions are considered. Tests for machine function on slopes are also used, and the total test series is intended to provide a total assessment of slope performance for a machine.

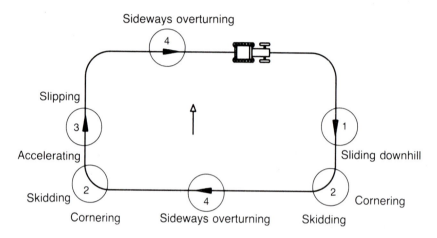

Fig. 4.13. Test hazards for four slope operations: (1) driving downhill, (2) cornering, (3) driving uphill, (4) driving on a side slope.

This approach has the merit of identifying all the potential limiting factors on slope operation for a particular machine, notwithstanding the simplified concepts which are used. Maler (1988) determined overall performance limits for a combine harvester of 14° on dry ground and 9° on wet ground, at speeds up to 6 km/h, using these tests. Carreel (1989) has written a computer program encompassing the test formulae, for evaluation in France.

Safe operating slope

Safe operating slope is difficult to define. In 1963 Molna proposed that the safe operating slope should be set 20° below the static stability limit, but this now looks impractical since the static limit for modern tractors has reduced to 33°, and trailed equipment may tip at 16°. The concept of using the static stability limit as a reference point is still valid.

Safe operating slope could well be defined as the slope where the probability of reaching the dynamic stability limit is low enough to be acceptable. This definition invokes criteria which can be evaluated by the methods of dynamics, statistics ('probability'), and economics ('acceptable'). The dynamic criteria have received the greatest attention, focusing on zero wheel load and overturning energy. Under dynamic conditions the zero wheel load criterion is not sufficient for overturning but it does define the operating point on a stability boundary. This in turn defines the total energy for overturning and hence the dynamic balance between potential energy and kinetic energy. First passage time is an appropriate statistical criterion for the process leading to overturning since it is a method of predicting failure due to random effects. These may arise from continuous changes in effective heading angle and effective slope caused by steering, acceleration and deceleration, as well as topography and rough ground. The economic criterion for what is acceptable, which in turn affects first passage time, has not yet been investigated.

Conclusions

1. Tractor overturning accidents remain a serious problem in agriculture. Non-fatal accidents are far more numerous than fatal accidents and cost approximately six times more to the national economy.

2. Overturning is the result of exceeding the limitations of the tractor. In an estimated 55% of cases the driver cannot judge when this will occur because he or she lacks information. The great majority of overturns are sideways, and are typically caused by slope, ground roughness, speed, acceleration, load and losing adhesion.

3. Slope puts a limit on performance as well as safety. On slopes exceeding 14°, grass becomes the only practical crop and most accidents occur on grass as a result. Increasing slope considerably reduces mechanization efficiency.

4. The safety cab remains the prime safety measure on tractors, but specialist mountain machines have also been developed and adaptations are being fitted on standard equipment for improved stability. Overturn prevention devices have not been successful but there is continuing interest in slope instrumentation for the cab.

5. Analysis of stability has now progressed from considering a special case, such as rearwards overturning, to determining an entire stability boundary which can be plotted on a polar diagram. This gives computed results for all the slope values, and heading angles on these slopes, at which wheel loads will reduce to zero.

6. The polar diagram can be used to display a very wide range of information. The partial stability limit at which a single wheel load will fall to zero can be compared with the total stability limit where a pair of wheel loads will fall to zero, and the effects of design changes can be superimposed on each other. The stability of a tractor can be compared with the stability of a trailer pulled behind it, and the effects of increasing the trailer load or fitting implements to the tractor can be shown. The minimum stability point for either tractor or trailer can be compared with the minimum point at which the machine combination will lose ground control.

7. The stability of articulated-steer machines is complex because of the altered geometry during steering, and because each alternative choice of position for the oscillation pivot results in different stability characteristics. In some designs the effect of steering can be to reduce the stability to a very low level. The stability of fluid-filled tankers is equally complex because the centre of gravity position is a function of fill, slope and heading angle. Minimum stability can occur when the tanker is only one-third full.

8. Dynamic effects can also be represented on the polar diagram by plotting the effective operating points in relation to the static stability boundary. The simplest effects are those of acceleration, either in a straight line or directed radially while cornering. The resulting shifts in effective operating locus on the polar diagram correspond to reduced stability margins at the relevant wheels. Whether the machine will ultimately overturn depends on whether the dynamic inputs are sustained for a sufficiently long time.

9. The effects of crossing obstacles on a side slope present the next level of complexity. The most commonly used overturning criterion is when the machine develops sufficient rotational kinetic energy to equal the required gain in potential energy. The dependence of overturning on slope, obstacle height and speed is demonstrated, and it is suggested that imminent overturning can be avoided by corrective steering or acceleration/deceleration. The difficulties of modelling tyre–ground interactions are regarded as the main source of divergence from experimental results.

10. Modelling the random effects of rough ground on slopes is likely to lead to the most valuable methods for defining safety from overturning. Estimating the first passage time to reach an overturning condition is regarded as more useful than determining the probability of falling within a range of extreme values. Experimental data have been used to estimate the largest values of rotational kinetic energy resulting from crossing different slopes on rough land.

11. The need to design strong safety cabs has prompted the dynamics during overturning to be modelled, and fully validated, with ground impacts and deformations of the tractor structure taken into account. The models make provision for multiple overturns before the tractor finally comes to rest.

12. Static stability can be measured using a tilt table, or calculated from standard test data on tractor geometry and centre of gravity position, but this is seldom done. An alternative method using portable equipment is now to be standardized, with the aim of encouraging every machinery manufacturer to measure and publish stability data. The method is based on weight-transfer measurements using weighpads on a moderate slope.

13. Dynamic stability tests using a standard obstacle have been developed, and incorporated into a series of tests on slope hazards and functions of a machine. Safe operating slope needs careful definition, however.

14. The principal research challenge which remains is to integrate the refined knowledge of static stability with a full understanding of dynamic behaviour on slopes, including random behaviour.

References

Baligand, J. (1978) Accidents at work to agricultural employees. *Bulletin* No. 250, CNEEMA, BP 121, 92164 Antony Cedex, France, November, pp. 37–45.

Biller, C. J. and Johnson, D. D. (1987) Apparatus for maintaining stability of mobile vehicles on sloping terrain. USDA Patent No. 4,679,803, US Patents and Trademark Office, Washington DC 20231, USA.

Blankenship, J. W., Means, K. H. and Biller, C. J. (1984) Side slope static stability of double-articulated logging tractor. Paper No. 841140, SAE, 400 Commonwealth Drive, Warrendale, PA 15096, USA.

Carreel, J. (1989) *Simulation of the Behaviour of Agricultural Machines on Sloping Ground*. CEMAGREF, Groupement de Clermont-Ferrand, Domaine des Pala-quins, Montoldre, 03150 Varennes-sur-Allier, France.

Chisholm, C. J. (1972) A survey of 114 tractor sideways overturning accidents in the UK–1969 to 1971. *Departmental Note* DN/TE/238/1425, National Institute of Agricultural Engineering, Wrest Park, Silsoe, Bedford.

Chisholm, C. J. (1981) Agricultural tractor overturning and impact behaviour. *Proceedings of the 7th IAVSD Symposium*, International Association for Vehicle System Dynamics, 7–11 September, Cambridge, UK, pp. 238–51.

Clark, S. (ed.) (1974) *Fundamentals of Machine Operation: Agricultural Machinery Safety*. John Deere Service Publications, Dept. F., John Deere Road, Moline, Illinois 61265, USA, pp. 12–15 and 146–60.

Coombes, G. B. (1968) Slope stability of tractors. *Farm Machine Design Engineering*, 18–33.

Crolla, D. A. and Spencer, H. B. (1984) Tractor handling during control loss on sloping ground. *Vehicle System Dynamics* 13, 1–17.

Daskalov, A. (1971) On the dynamic stability of tractors against overturning. *Selskostopanska Tekhnika* 8 (5), 3–14.

Daskalov, A. (1972) On the effect of agricultural machines on the dynamic resistance of tractors to overturning. *Selskostopanska Tekhnika* 9 (2), 15–25.

Daskalov, A. (1973) The effect of the movement parameters on the dynamic resistance of tractors to overturning. *Selskotopanska Tekhnika* 10 (6), 33–47.

Daskalov, A. (1974) On some properties of the limiting equilibrium equation with the studies of dynamic resistance of tractors to overturning. *Selskotopanska Tekhnika* 11 (3), 33–46.

Daskalov, A. (1986) The influence of the main inertial moments on the dynamic stability of tractors. *Selskotopanska Tekhnika* 23 (6), 74–80.

Davis, D. C. and Rehkugler, G. E. (1974a) Agricultural wheel-tractor overturns, Part 1: Mathematical model. *Transactions of the ASAE*, 477–83, 492.

Davis, D. C. and Rehkugler, G. E. (1974b) Agricultural wheel-tractor overturns, Part 2: Mathematical model verification by scale-model study. *Transactions of the ASAE*, 484–92.

Dettwiler, E. and Gammenthaler, H. U. (1975) Mechanisation of hill farming. *Schweizerische Landwirtschaftliche Forschung* 14 (4), 371–93.

Dillman, E. R. (1973) Vehicle overturn preventer. US Patent No. 3712405.

Feng, Y. and Rehkugler, G. E. (1986) A mathematical model for simulation of tractor sideways overturns on slopes. Paper No. 86-1063, ASAE, St Joseph, MI 49085-9659, USA.

Finlayson, D. M. (1989) Experimental validation of a computer simulation of the static stability of articulated steer vehicles, using a scale model: Phase 1. *Dep. Note 27*, Scottish Centre of Agricultural Engineering, Penicuik, EH26 0PH.

Finlayson, D. M. (1990) A computer model to predict the static stability of an articulated steer vehicle having a rigid body. *Dep. Note 31*, Scottish Centre of Agricultural Engineering, Penicuik, EH26 0PH.

Gibson, H. G. and Biller, C. J. (1974) Side-slope stability of logging tractors and forwarders. *Transactions of the ASAE* 17, 245–50.

Gibson, H. G., Elliot, K. C. and Persson, S. P. E. (1971) Side slope stability of articulated-frame logging tractors. *Journal of Terramechanics* 8 (2), 65–79.

Gibson, H. G., Thorner, B. C. and Thomas, J. W. (1981) Slope stability warning device for articulated tractors. US Patent – 4 284 987, Department of Agriculture, Washington DC, USA.

Gillham, M. H. (1985) A program to calculate stability diagram co-ordinates. *Dep. Note* no. SIN/424 (revised), Scottish Institute of Agricultural Engineering, Penicuik, EH26 0PH.

Goldberg, J. H., Parthasarathy, V. and Murphy, D. J. (1989) Operator limitations in tractor overturn recognition and response. Paper No. 89-1107, ASAE, St Joseph, MI 49085-9659, USA.

Grace, L. S., Biller, C. J., and Means, K. H. (1983) A survey of tractor stability analysis. Paper No. 83–1618, ASAE, 2950, Nile Road, St Joseph, MI 49085, USA.

Grecenko, A. (1983) Dynamic stability as a component of the gradeability of farm machines. *Zemedelska Technika* 29 (11), 643–58.

Grecenko, A. (1984a) Operation on steep slopes: state-of-the-art report. *Journal of Terramechanics* 21 (2), 181–94.

Grecenko, A. (1984b) Study of the motion of agricultural vehicles on steep grass-covered slopes. *Proceedings of the 8th International Conference of the International Society for Terrain-Vehicle Systems*, Cambridge, England, 6–10 August, pp. 595–614.

Grecenko, A. (1986) A method of assessing the slope performance of agricultural vehicles. *Zemedelska Technika* 32 (10), 577–98.

Health and Safety Executive (1986) *Agricultural Black Spot: A Study of Fatal Accidents.* HSE, St. Hugh's House, Bootle.

Hunter, A. G. M. (1979) Characterisation of rough ground using an accelerometer for measurement. *Journal of Terramechanics* 16 (1), 33–44.

Hunter, A. G. M. (1981) Critical direct descent and ascent slopes for an agricultural tractor with forage harvester and trailer. *International Journal of Vehicle Design* 2 (3), 289–98.

Hunter, A. G. M. (1982a) Tip angles for tractor sideways overturning from Norwegian test reports. *Dep. Note* No. SIN/355, Scottish Institute of Agricultural Engineering, Penicuik, EH26 0PH.

Hunter, A. G. M. (1982b) A physical model for demonstrating tractor accidents. *Journal of Agricultural Engineering Research* 27, 163–8.

Hunter, A. G. M. (1984) Some stability and control problems with trailed farm tankers on slopes. *Journal of Terramechanics* 21 (3), 273–82.

Hunter, A. G. M. (1985a) A theoretical analysis of the stability of a Heywang 10/25 self-loading forage wagon. Confidential report, Scottish Centre of Agricultural Engineering, Penicuik, UK.

Hunter, A. G. M. (1985b) Stability analysis of trailed tankers on slopes. *Journal of Agricultural Engineering Research* 32, 311–20.

Hunter, A. G. M. (1986) Centre of gravity analysis of fluid in inclined tanks. *Journal of Agricultural Engineering Research* 33, 111–26.

Hunter, A. G. M. (1988) Stability of a prototype backhoe loader: computer study. Confidential report, Scottish Centre of Agricultural engineering, Penicuik, UK.

Hunter, A. G. M. and Owen, G. M. (1983) Tractor overturning accidents on slopes. *Journal of Occupational Accidents* 5, 195–210.

Hunter, A. G. M., Owen, G. M. and Glasbey, C. A. (1988) Safety on steep slopes: Dynamic stability experiments. In: Cox, S. W. R. (ed.), *Engineering Advances for Agriculture and Food: Proceedings of the 1938–1988 Jubilee Conference of the Institution of Agricultural Engineers*, 12–15 September, Cambridge, pp. 229–30.

Hunter, A. G. M., Owen, G. M. and Glasbey, C. A. (1990) Review of experimental results, analysis of prediction accuracy, and final assessment of weighpad method. Final report (confidential) DTI Ref RTP2/155/70, Scottish Centre of Agricultural Engineering, Penicuik, UK.

Jensen, D. V. (1980) A summary of fatal farm accidents in Wisconsin from 1944 to 1978. In: *Engineering a Safer Food Machine*, ASAE, 2950 Niles Road, PO Box 410, St. Joseph, MI 49085, USA, pp. 120–6.

Kaloyanov, A., Sredkov, S. and Stoichev, S. (1974) On the static stability of wheeled tractors. *Selskostopanska Tekhnika* 11 (7), 45–57.

Kelly, J. E. and Rehkugler, G. E. (1980) Stability criteria for tractor operation on side slopes. In: *Engineering a Safer Food Machine*, ASAE, 2950 Niles Road, PO Box 410, St Joseph, MI 49085, USA, pp. 145–57.

Kim, K. U. and Rehkugler, G. E. (1987) A review of tractor dynamics and stability. *Transactions of the ASAE* 30 (3), 615–23.

Larson, D. and Liljedahl, J. B. (1971) Simulation of sideways overturning of wheel tractors on side slopes. Paper No. 710709, Society of Automotive Engineers, Inc., Two Pennsylvania Plaza, New York, NY 10001, USA.

Liljedahl, J. B., Carleton, W. M., Turnquist, P. K. and Smith, D. W. (1979) *Tractors and their Power Units*, 3rd edn. John Wiley and Sons, New York.

McKibben, E. G. (1927) The kinematics and dynamics of the wheel-type farm tractor. *Agricultural Engineering* 8 (1–7), 24 pp.

MAFF and DAFS (1969) *Farm Safety: Guide to the Safety, Health and Welfare Act and Regulations*. MAFF, Whitehall Place, London.

Maler, J. (1988) The gradeability of standard combine harvesters. *Zemedelska Technika* 34 (8), 469–79.

Manby, T. C. D. (1963) Safety aspects of tractor cabs and their testing. *Journal and Proceedings of the Institution of Agricultural Engineers* 19 (3), 53–68.

Marples, V. (1971) An analysis of tractor accidents in England and Wales. *Occupational Safety and Health*, 1 Jan, 19–23.

Meyer, H. (1956) Tyre tests with self-aligning tractors on inclined surfaces. *Landtechnische Forschung* 5, 139–42.

Molna, B. (1963) The tractor accidents and factors increasing or decreasing the risk of such accidents. Forsoksmelding 8, Landsbruksteknisk Institutt, Vollebekk, Norway.

Monk, A. S., Morgan, D. D. V., Morris, J. and Radley, R. W. (1986) The cost of accidents in agriculture. *Journal of Agricultural Engineering Research* 35, 245–7.

Murphy, D. J. and Johnson, S. R. (1984) Tractor overturns: a new preventative approach. *Agricultural Engineering* 65 (1), 15–17.

Murphy, D. J., Beppler, D. C. and Sommer, H. J. (1985) Tractor stability indicator. *Applied Ergonomics* 16 (3), 187–91.

Newland, D. E. (1981) The roll stability of frame-steered two-axle vehicles during steady cornering. *Proceedings of the 7th IAVSD Symposium*, International Association for Vehicle System Dynamics, 7–11 September, Cambridge, UK, pp. 252–66.

OECD (1988) *OECD Standard Codes for the Official Testing of Agricultural Tractors*. OECD Publications Service, 2 Rue Andre-Pascal, 75775 Paris Cedex 16, France.

Ott, A. (1978) Mechanisation of forage crops on slopes. *Technique Agricole* 8.

Ott, A. (1989) Current state and development of mechanisation in mountain regions. *Technique Agricole* 12, 10–14.

Owen, G. M. (1980a) The effects of offset load and travel speed on tractor stability. *Dep. Note* SIN/317, Scottish Institute of Agricultural Engineering, Penicuik, UK.

Owen, G. M. (1980b) The effects of radial and cross-ply tyres on tractor stability. *Dep. Note* SIN/318, Scottish Institute of Agricultural Engineering, Penicuik, UK.

Owen, G. M. (1981) The effect of a mounted hay tedder on tractor stability. *Dep. Note* SIN/316, Scottish Institute of Agricultural Engineering, Penicuik, UK.

Owen, G. M. (1986) Two types of tractor four-wheel drive engage alarm for use on slopes. *Journal of Agricultural Engineering Research* 33, 155–8.

Owen, G. M. (1987) Trailed equipment stability on slopes: problems and solutions. *Agricultural Engineer*, 124–7.

Owen, G. M. and Hunter, A. G. M. (1983) A survey of tractor overturning accidents in the United Kingdom. *Journal of Occupational Accidents* 5, 185–93.

Owen, G. M. and Hunter, A. G. M. (1987) A safe slope monitor for agricultural tractors: survey of use on farms. *Proceedings of the 9th International Conference of the International Society for Terrain Vehicle Systems*, 31 August to 4 September, Barcelona, Spain, pp. 743–51.

Owen, G. M. and Hunter, A. G. M. (1988) A microprocessor-based vehicle static stability limit calculator for field use. *Journal of Agricultural Engineering Research* 40, 233–6.

Owen, G. M. and Spencer, H. B. (1980) Overturning and control loss experiments with a radio-controlled tractor. *Dep. Note* SIN/319, Scottish Institute of Agricultural Engineering, Penicuik, UK.

Pershing, R. S. and Yoerger, R. R. (1969) Simulation of tractors for transient response. *Transactions of the ASAE* 12 (5), 715–19.

Quigley, A. D. (1986) 'A study of tractor tyre braking on agricultural land.' Ph.D. Thesis, Heriot-Watt University, Edinburgh, UK.

Rehkugler, G. E. (1978) Simulation of articulated steer four-wheel drive agricultural tractor motion and overturns. *ASAE Technical Paper* no. 78-1030, ASAE, St Joseph, MI 49085, USA.

Reichmann, E. (1972a) Slope stability of agricultural vehicles, Part 1: Tip boundaries for three- and four-wheel vehicles driven on a plane slope. Heft 1, *Forschungsberichte der Bundesversuchs- und Prufungsanstalt für landwirtschaftliche Maschinen und Gerate*. Wieselburg, Austria.

Reichmann, E. (1972b) Stability criteria for agricultural vehicles driven on slopes. In: Scheruga (ed.), *25 Jahre, Bundes-Versuchs- und Prufungsanstalt für landwirtschaftliche Maschinen und Gerate*. Wieselburg, Austria, pp. 37–45.

Ribetou, P. (1984) Contribution to improving safety for forage harvesting in the mountains. *Memoire de fin d'études*. CEMAGREF, Groupement de Clermont-Ferrand, Echelon de Montoldre, 03150 Varennes-sur-Allier, France.

Schneider, R. D. (1980) The Nebraska farm accident example. In: *Engineering a Safer Food Machine*, ASAE, 2950 Niles Road, PO Box 410, St Joseph, MI 49085, USA, pp. 17–23.

Schwanghart, H. (1971) Tractor overturning and testing of safety devices against overturning. *Landtechnische Forschung* 19 (1), 1–5.

Schwanghart, H. (1973) Calculation method for prevention of continuous overturning of agricultural tractors on a slope. *Grundlagen der Landtechnik* 23 (6), 170–6.

Schwanghart, H. (1978) Tipping and overturning behaviour of tractors. *Landtechnik* 11, 488–92.

Sestak, J., Skulavik, L., Sklenka, P. and Markovic, R. (1989) Dynamic stability of a general purpose field power unit for mountain areas. *Zemedelska Technika* 35 (10), 579–96.

Spencer, H. B. (1978) Stability and control of two-wheel drive tractors and machinery on sloping ground. *Journal of Agricultural Engineering Research* 23, 169–88.

Spencer, H. B. (1988) *Static Stability of Articulated Steer Tractors*. Scottish Centre of Agricultural Engineering, Penicuik, EH26 0PH.

Spencer, H. B. and Gilfillan, G. (1976) An approach to the assessment of tractor stability on rough sloping ground. *Journal of Agricultural Engineering Research* 21, 169–76.

Spencer, H. B., Owen, G. M. and Greenhill, A. L. (1983) A microprocessor-based safe descent slope meter. *Journal of Agricultural Engineering Research* 28, 269–72.

Spencer, H. B., Owen, G. M., and Glasbey, C. A. (1985) On-site measurement of the stability of agricultural machines. *Journal of Agricultural Engineering Research* 31, 81–91.

Van Deusen, B. D. (1967) *A Statistical Technique for the Dynamic Analysis of Vehicles Traversing Rough Yielding and Non-yielding Surfaces*. NASA CR-659, National Aeronautics and Space Administration, Washington DC, USA.

Whitaker, J. R. (1973) An analysis of overturning accidents with safety cabs. Paper No. 6, NIAE Subject Day 'Tractor Ergonomics', National Institute of Agricultural Engineering, Silsoe, Bedford MK45, UK.

Willsey, F. R. and Liljedahl, J. B. (1969) A study of tractor overturning accidents. Paper No. 69-639, ASAE, PO Box 229, St Joseph, MI 49085, USA.

Wooley, J. C. (1921) Levelling device for tractors working on hillsides. *Agricultural Engineering*, p. 227.

Wray, G., Nazalewicz, J. and Kwitowski, A. J. (1984) Stability indicators for front end loaders. *Proceedings of the 8th International Conference, International Society for Terrain-Vehicle Systems*, Cambridge, UK, 6–10 August.

Zakharyan, E. B. (1972) Evaluation of the dynamic overturning stability of a tractor and trailer outfit from the energy point of view. *Traktory i Selkhozmashiny* 7, 2–4.

Zwaenepoel, P., Beaulieu, G. and Boffety, D. (1987) Risks of loss of control on tractors descending grass fields – influence of tyre drive factors. BTMEA no. 17, CEMAGREF, Groupement du Clermont-Ferrand, Echelon de Montoldre, 03150 Varennes-sur-Allier, France.

Chapter 5

The Principles of Robotics in Agriculture and Horticulture

Francis Sevila and Pierre Baylou

Robots: A new age for agricultural machinery

New types and systems of agricultural machinery are being developed under the influence of various social, economic and technical factors.

1. The entrepreneurial farmer is now becoming the norm; his or her role has diversified from being purely a food producer to that of nature and environment protector.

2. The performance of electronic components and software development methods are constantly improving, while their prices continue to fall. The potential computing power that could equip a tractor in the late 1990s may well be equivalent to that of a present day computing centre.

3. Industry and laboratories specializing in automation are now joining their efforts to agricultural machinery manufacturers. The latter are developing their skills and know-how in technologies that they were not using until very recently. Consequently, machines for agriculture will be more and more of the robotic type, which means that they will include a growing proportion of automation and sensor technology for their executive and mobility functions (Lucas, 1984; Marchant, 1985; Ito, 1990).

Definition of an agricultural robot

Many agricultural machines are already partly automated, performing repetitive tasks when set up by human operators. In industry, the term 'robot' is dedicated to machines operating without direct human involvement on a wide variety of tasks and objects. They include:

- sensors;
- data-processing systems; and
- actuators for adaptative and selective actions.

Whenever an agricultural machine has a complex sub-system that can be described in the above terms, it can be called a robotic system.

Specifications for agricultural robots

Robotic systems are related to the type of task, to their environment, and to the user and production conditions. These factors lead the designers of robots for agriculture to specific constraints compared to robots for factory use. Designers have devised original methods and systems in order to meet these constraints. These can be characterized in the following order:

1. Detection of objects with a significant biological variability and in complex biological environments, under, in some cases, natural lighting conditions.
2. Interactive operations between complex sensors and actuators in continually changing and not easily modelled scenes and tasks.
3. Modelling of biological objects for the design and control of the corresponding robotized systems.
4. Control of robotized structures adapted to all-terrain, mobile and hostile conditions.
5. Design of specific actuators and end-effectors adapted to various constraints, including the fragility of the biological objects they have to handle, and the speed, precision, maintenance, and price constraints for their agricultural uses.
6. Multi-sensor mountings using devices of various physical principles, precisions and reliabilities.
7. Simulations of human perceptions and decisions on complex objects and tasks for real-time and on-machine use.

Detection in complex environments

In the case of crops growing in open fields, the processing robot has to withstand bad weather, dust, a variety of obstacles, and other adverse conditions. The ground can also present different aspects: for example, slopes, irregular surfaces, and the plants of the crop can vary significantly. In an industrial environment robots are not usually beset by so many constraints and variabilities. But the agricultural environment is so unpredictable that the robot needs to be equipped with, for example, a redundant multi-sensor system in order to preserve its integrity as well as avoid any damage caused by malfunction.

Objects that have to be handled generally have a very poor accessibility. Fruits can be difficult to detect and reach since they are often hidden by leaves and positioned among branches (Rabatel, 1988; Slaughter and Harrell, 1989). Certain vegetables, like white asparagus, have to be picked before emerging into the open air; they have to be detected as soon as they reach the soil surface (Ouali, 1986).

Among the various sensors that have been put into operation for detection and control purposes, machine vision has gained a very privileged role.

The privileged role of vision

Technical and economic aspects

Visual sensors tend to be the type best suited for observing the wide range of sizes, shapes and colours encountered in agriculture (Tutle, 1983; Bourely *et al.*, 1986; Rehkugler and Throop, 1986; Davenel *et al.*, 1988b; Grand d'Esnon *et al.*, 1988; Rigney and Kranzler, 1988; Marchant *et al.*, 1989; Slaughter and Harrell, 1989; Bennedsen *et al.*, 1990; Brown *et al.*, 1990; Ding *et al.*, 1990; Han and Hayes, 1990; Howarth *et al.*, 1990; Humburg and Reid, 1990; Ling *et al.*, 1990; MacDonald and Chen, 1990; Mechineau *et al.*, 1990; Stafford and Ambler, 1990; Tao *et al.*, 1990; Tohmaz and Hassan, 1990). For this reason, more and more attention is paid to cameras, with increasingly better performance and lower prices. Moreover, cameras are becoming smaller and more robust, making them more attractive for incorporating in mobile field machines and farmstead equipment.

However, problems are encountered with the large amount of data to be processed from these cameras, which may approach 1 to 2 Megabytes per image frame for colour cameras. This difficulty has to be overcome in order to achieve real-time operation.

Data processing

One way to solve this problem is to use adapted computer architecture, like vectorial processors, or microprocessors specially devoted to signal processing (DSP) (Teneze *et al.*, 1989). However, processing time is often related to the 'complexity' of the image. This complexity varies with the definition of the camera, the number of dimensions (one, two or three) of the geometry of the scene, the light dynamic (for example, diffuse uniform light, shade with sunlight spots), and whether the case involves monochrome or colour information.

Visual sensor definition

From the simple optical barriers (some tens of diodes) used to detect the opening of crocus flowers, or the arrival of a cow's teats in the milking cups, to the 25,000 diodes of the CCD cameras used for fruit detection, the range of possible tools is wide. For each problem a visual sensor can be specifically chosen, minimizing the number of useful light beams required.

The geometry of the problem

Image simplification can be achieved by adjustment of the optics and through the methodology. For example, the positioning in three dimensions of the teats of a cow for robot milking can be obtained by an appropriate detection of the monochromatic light segments of a sweeping laser beam reflected by these teats (Mechineau *et al.*, 1990; Fig. 5.1).

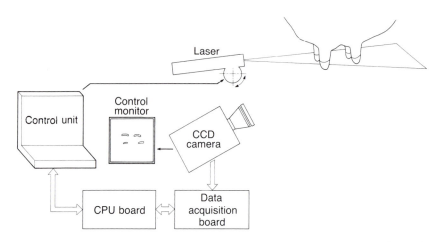

Fig. 5.1. A vision system for a cow milking robot. (CEMAGREF, Antony, France.) The system uses a flat beam of laser light that sweeps up and down the udder, and a CCD camera placed at a suitable position and distance. Segments of light are reflected into the camera and then processed. The three-dimensional structure of the udder can therefore be constructed by appropriate software that uses the successive one-dimensional segments generated by the laser's sweep.

In the case of vine-branch detection for winter pruning, the problem is transformed from a three- to a two-dimensional one by two complementary actions.

1. An 'edging' action on the vine row drastically reduces the dimension of the vine perpendicular to the row. The vine can then be approximated as being of a negligible size in this direction.

2. A back-lighting of the vine towards the camera along this reduced dimension allows the grabbing of a two-dimensional image containing most of the necessary information (Sevila, 1985).

It is sometimes sufficient to examine the contour of objects: in the case of white asparagus detection, a combination of back-lighting and of contour analysis simplifies a three-dimensional problem into a one-dimensional one: the measure of the height of the ground line (Ouali, 1986; Fig. 5.2).

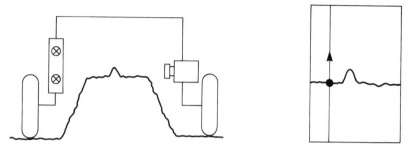

Fig. 5.2. The robotic harvesting of white asparagus. (ENSERB, Bordeaux.) Diascopic lighting is placed on one side of the row and a camera on the other. The original three-dimensional problem is thereby translated into a one-dimensional one: the detection of an asparagus tip above the ground line.

Light dynamics

Large variations of light, such as those encountered by machines working in the open fields, make image processing and feature extraction much more difficult. Various attempts have been made to reduce light variations, including:

● *shading measures* to control the sunlight, using straddle frames or umbrella-type devices. In these situations the objects of interest can also be isolated from the surroundings, using uniform background screens; and
● *artificial lighting* using continuous or intermittent light (diffuse, specular, stroboscopic, flash, polarized and so on). The intensity of sunlight is difficult to match during an all-day activity, especially when the light is at a low incident angle.

The design of such lighting solutions is closely related to the choice of the corresponding camera and image processing software.

Spectral situation

When there is a clear colour contrast between the object to be detected and its background – for example, coloured mature fruits in their green foliage

− it may be possible to reduce the detection difficulties related to the natural lighting dynamic by using multi-spectral analysis.

Let $F_o(w)$ and $F_b(w)$ be the spectral reflectivity of the object and its background, versus the wavelength w of the light. If two wavelengths w_1 and w_2 can be found such that:

- F_o is high and F_b is low for w_1, and
- F_o and F_b are similar, or F_o is low and F_b is high for w_2,

then the ratio

$$F_o(w_1)/F_b(w_2)$$

will be high on object pixels (picture elements of the detected image) and low on background pixels. This ratio is almost independent of light intensity whenever the corresponding pixels are working in their range of light intensity, and are not saturated (Rabatel, 1988; Slaughter and Harrell, 1989; Molto *et al.*, 1990).

Interactive operations between sensors and actuators

The complexity of the sensors, actuators and their working environment have until now led agricultural robot designers to develop their own concepts specifically adapted to each problem to be solved. In contrast, many industrial robots have been developed through the use of already available sensing and actuating arrangements.

For agricultural robots, one of the problems to be solved is the global calibration of the sensors arrangement versus the mechanical parts adjustment. For instance, for vision-guided agricultural robots, special calibration methods had to be designed in order to obtain rapid on-site recalibration of the system (Tillett, 1989; Ling *et al.*, 1990).

Apart from the special arrangements of frames and manipulators (the robot's arms and end effectors, and autonomous mobile frames; see below), efforts have been concentrated on the parts of the agricultural robots that are in working interaction with the living plant or animal: the sensors, the actuators, and its end effector (Sittichareonchai and Sevila, 1989; Ferrand *et al.*, 1990; Simonton, 1990b; Wolf *et al.*, 1990).

Since global detection devices are not sufficiently accurate for complete control of the tasks, actuators are often used to complete the sensing operation by providing a local collection of information in the plant foliage, or close to the animal or the ground. Examples include: machine−furrow distance-sensing completing the absolute field positioning of a ploughing robot (Bonicelli, 1987); the on-the-cluster array of light beams which helps in precisely positioning the milking cup on the cow teat (Mechineau, 1990;

Fig. 5.6); the air depression detector which helps to stop the motion of the picking arm towards the fruit as soon as the fruit enters the suction cup of the clamp (Bourely *et al.*, 1990; Fig. 5.5.); the strain-gauge arrangement optimizing the position of the wool-cutting shear along the skin of the sheep (Key, 1985); and the camera mounted on the end-effector (pruning shear or picking clamp) in order to precisely and dynamically position it before action (Monsion, 1985; Harrell *et al.*, 1990b).

Modelling of biological objects

As for industrial applications, living objects to be handled or transformed by agricultural robots need to be modelled through descriptions using a limited and fixed list of geometrical and physical parameters. Mathematical models have been developed in order to simulate the behaviour, growth, and shapes and structures of plants and animals.

The life sciences, especially when dealing with agricultural matters, were areas where numerical simulation was rarely used before the need arose for such simulation for the automation of control or classification.

The various models can be classified into the following different types.

1. Those using real botanical or animal characteristics obtained by statistical measurements. Such models use global parameters, like global volume of foliage, density or vegetation, and trunk–branch angles (Giles *et al.*, 1988; Grand d'Esnon *et al.*, 1988; Bourely *et al.*, 1990; Howarth *et al.*, 1990; Jia and Krutz, 1990; Marchant and Schofield, 1990; Tao *et al.*, 1990).
2. Those of the computerized type, which totally represent the plant by mathematical objects having properties or structures similar to the represented plant or animal (Sevila, 1985; Sevila *et al.*, 1990).

Most of the shapes and processes that are modelled in such programs include random characteristics, due to the natural variety of living objects. The construction of such models implies a multidisciplinary team of engineers, plant or animal scientists and computer specialists.

Robotic vehicles

The main problems that have been worked on for robotic mobile devices in agriculture are guidance and inclination control. The corresponding parameters of the robot are: *guidance* (see Fig. 5.4), including position (X_r, Y_r) relative to the coordinate axis, and attitude (angle T_r) relative to the direction of travel; and *inclination*, including roll (R_r) and pitching (P_r) angles related to vertical planes.

For some agricultural tasks it is also necessary to know the altitude Z_r: for example, for machines for drain installation the altitude has to be known very precisely in order to secure an accurate and regular drain slope. Usually, however, an agricultural machine does not need to know its Z_r coordinate to control its mobility.

Guidance: the special case of agriculture

Types of working areas

Many types of environments may be found in agriculture. They mainly belong to one of the following types:

1. *Void areas*, where the working space does not have any object detectable by the robot sensors, such as a field to plough.
2. *Populated areas,* with either:
   (a) *structured populations*, either (i) with one-dimensional periodicity, like corn and tobacco (Reid, 1988); or (ii) with two-dimensional periodicity, like vines and orchards.
   or
   (b) *unstructured populations* of randomly situated objects, like forests (Bonicelli *et al.*, 1989).
3. *Work-structured areas*, where the work performed creates a clearly detectable frontier between processed and unprocessed areas, such as in a wheat field in front of a combine harvester, or on a lawn in front of a mower.

The agricultural task and path

The automated mobility function requires the use of perceptual systems composed of sensors and data processors that control the vehicle's movements. The agricultural task undertaken defines the path the robot has to follow. This path may be determined through: (i) objects normally in the working area, such as trees, rows, furrows; (ii) objects installed and artificial phenomena created there such as beacons, laser beams, radar emitters and reflectors; and (iii) external information, for example, from a satellite or a modelled trajectory stored in the computer memory.

In contrast to the study of military or public transport robots, no published work has treated the case of an agricultural mobile robot constructing its path with wide opportunities of choice according to its strategy of action and complete perception of its environment. Automated path-search is usually limited for agricultural robots to simple transit problems, such as finding the next line of apple trees to be picked, the next furrow to plough or the next field lane to spray.

In-field all-terrain location

This modest state of the art is related to the location difficulty in agricultural conditions, as well as in military conditions. The main problem that remains in automatic mobile control is to know where the robot is actually located in an often poorly structured environment, with a wide variability of characteristics.

Standard agricultural locating problems are more difficult to solve than comparable ones in the manufacturing industry. The differences are based on: (i) area of movement (agriculture − 1 km^2; industry − 0.01 km^2); (ii) nature of the ground (agriculture − hilly, muddy, uneven; industry − flat, regularly structured and concreted); and (iii) machine power (agriculture − 100 kW; industry − 1 to 10 kW).

The first point implies, for instance, that the use of an electrical guiding wire placed deep into the ground is generally difficult. Moreover, vermin are likely to rapidly damage such wires. The second implies trajectory error sources much bigger in the agricultural case, especially on slopes. Finally, safety problems, and consequently guidance quality, are much more crucial when the mobile robot is a very powerful one.

Various locating methods

Possible locating methods belong to one of three basic types (Young *et al.*, 1976): absolute, relative, and iterative. Each will be described in more detail below.

Absolute location

This type of location occurs when the robot position is known in relation to stable and well-identified axes of reference (Monod and Mechineau, 1986; Mechineau, 1988; Fig. 5.3). There are four possible cases:

1.   Direct detection of objects belonging to the reference system, such as beacons and cables (Shmulevich *et al.*, 1989; Tillett and Nybrant, 1990).
2.   Indirect detection of objects in the working area, whose absolute position is known (Choi, 1990).
3.   Non-physical structuring of the work area, by gravity or magnetic fields, or by artificially generated electromagnetic waves (lasers, radio frequency) (Palmer, 1990).
4.   Use of readings of an on-machine receiver of satellite information (Buschmeier and Müller, 1990).

Relative location

Global axes of reference are not obligatory. By locally sensing the environment of the robot, enough information may be collected to perform

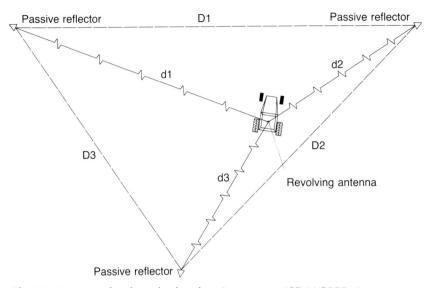

Fig. 5.3. An example of an absolute location system. (CEMAGREF, Antony, France.) A revolving radar-type antenna mounted on the vehicle sweeps the whole field in which three tetrahedral reflectors are installed at known fixed positions. The distances d1, d2 and d3 are computed from the radar information, and when related to the reference distances D1, D2 and D3 provide an absolute coordinate system for the mobile machine.

the task (Bonicelli and Monod, 1987; Defranco, 1988; Reid and Searcy, 1988). Various cases are found in robots for agriculture. Examples include: (i) plough furrow following by a ploughing robot; (ii) mound-of-earth following in asparagus picking; (iii) row of trees in apple picking (Fig. 5.4.); and (iv) the main direction of row crops.

For relative location, the object to be sensed is either continuous or discrete. Procedures have to be implemented to take into account accidental discontinuities, such as missing trees in the row for a picking robot. When robot control can use such route guides as furrows, lines of trees, rows of plants (but also wires or laser beams), the complex guidance problem may be reduced to two partial and simpler location problems: (i) following the path defined by the route guide; and (ii) measuring the distance covered along the path. The second parameter does not always have to be known with precision, but in some applications, it is necessary to know the exact longitudinal position along the path.

Such precision is necessary when planting grape vines with an automatic machine, if the vine plants are to be seen well-aligned from all directions. Steering the machine to follow a straight row may be achieved by relative guidance following an already ploughed furrow, or a trajectory based on

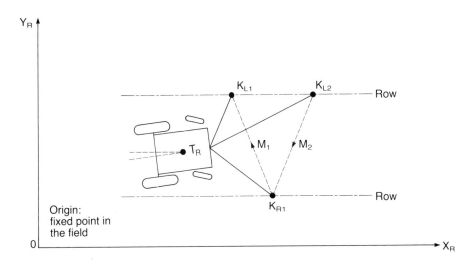

Fig. 5.4. Relative location. (ENSERB, Bordeaux.) The robot is moving between rows; within its sensor range it detects objects K_{L1}, K_{L2} on the left and K_{R1} on the right. The robot can then compute its next positional targets M_1 (half-way between K_{L1} and K_{R1}) and M_2 (half-way between K_{L2} and K_{R1}). M_1 and M_2 lie along the centre-line between the rows, and thus show the robot the path it must follow. By contrast, absolute location would use the independent coordinates X_R, Y_R and T_R instead.

a planar laser beam. A longitudinal guidance system is needed to plant at very regular and exact distances along the row. A simple in-field absolute location is made by means of string that unrolls as the machine advances. The on-board measurement of the string length unravelled indicates to the control system the longitudinal position of the machine along the planted row.

Iterative location

In this case the robot proceeds by consecutive position estimations using iterative algorithms. At each step the position is computed from the previous step position, to which a measured increment is added. This increment is obtained by internal or external measurements made by the robot sensing devices.

For *internal measurements*, the control system may take its data from the robot internal sensing units. These could include a pedometer (Ollila *et al.*, 1990), accelerometer or gyroscope. The main problem encountered with this method is the accumulation of errors that may need to be regularly corrected by absolute relocations.

The pedometer is the most widely used device for iterative type location: robot wheel rotation increments are counted and integrated over time. Slippage on muddy ground is often the main problem with such methods, since the wheel rotation does not exactly correspond to the advance of the robot. An external measurement is needed: ground speed is generally obtained with a radar system using the Doppler effect.

In the case of *external measurements*, when the task has a structuring effect on the working area, the detectable component of this structure is used for guidance purposes as, for example, in tillage and wheat harvesting (Brown *et al.*, 1990).

In such cases a clearly defined frontier exists between the field where the work has not been done and the one with the work done. Although similar to a relative location method, this situation is iterative since position reading by the machine is dependent on another reading made on the previous pass. Such positioning might suffer from the quality of the actual task (such as the quality of the cutting frontier).

Special beams can also be distributed while the machine is advancing to ensure vehicle guidance during the next pass. The accumulated errors in this latter step-wise location method can be low.

Inclination control

Machine chassis position control relative to verticality or ground profile is obtained by internal or external sensing devices.

Verticality is sensed with gravity field sensing devices, the most common one being based on position readings of a damped pendulum attached to the machine. These readings are often not sufficient to obtain rapid dynamic control of chassis verticality. Accelerometric devices reading instantaneous values of the machine chassis rotational acceleration are therefore necessary. Mounted in parallel with pendulum type sensors they can lower the basic verticality control cycle time from 2 s to 0.2 s for a full-size tractor (Rabatel and Bonicelli, 1986).

Machine chassis position relative to the ground profile is obtained through distance measurement from parts of the frame to the closest ground area. Ultrasonic distance sensors are convenient for this purpose since the distance to be measured is between 0.2 and 2 m (Marchant, 1985; Monod and Bonicelli, 1986).

Design of specific actuators

For each task, actuators (robot arms) have to be provided. Because they have to follow specific constraints due to their agricultural use, they are seldom direct copies of industrial types of actuators (Guul-Somonsen, 1990).

Constraints deal on the one hand with low selling cost possibilities, random shocks and accidents, and user experience and education; and on the other hand with characteristics of the agricultural tasks to be performed. These factors lead to design constraints for actuators in agricultural robotics (Chen and Holmes, 1990; Ferrand *et al.*, 1990; Simonton, 1990b).

Design constraints on actuators

Velocity requirements

Cycle times of 1 to 2 s are often encountered for operating cycles whose characteristics may vary significantly from one cycle to the next.

For instance, a fruit picking arm (Bourely *et al.*; Fig. 5.5) has an active area of $1 \, m^2$ in front of it, and an active depth of $0.60 \, m$. It has to pick fruit

Fig. 5.5. Magali, a robot for harvesting apples. (CEMAGREF, Montpellier.) While the robot advances between adjacent rows of fruit trees, its vision systems detect the fruits on both sides. Picking arms then gently pick the fruits and transfer them to automatic conveying and storage devices.

at any place in this volume. A special effort was made to alternate picking from one side of this working volume to the other one. This allows the fruit disturbed on the first side by the previous detachment action to come to rest before the next picking. The method has a significant effect on the total picking efficiency, since a moving fruit is more difficult to detect, position and pick. But with a basic constraint of 2 s per cycle, such a picking strategy implies a very rapid arm movement and hence a high actuator velocity.

Low precision

When compared to industrial robotic actuators, precision in the agricultural environment can be much lower (Bourely, 1989). Because of the variability of the dimensions and position of objects, the end-effectors have to be adaptative enough to provide the robot with the necessary compliance (Sittichareonchai and Sevila, 1989; Wolf *et al.*, 1990). Current precision specifications are: *static*, 1 mm to 1 cm instead of 0.1 mm; and *dynamic*, centimetres instead of millimetres. This has a direct consequence on the type of mechanisms, sensors and electronic components that can be used for agricultural robot actuators: they can be less precise and less costly.

Powering of agricultural robot arms

The velocity constraint has a direct effect on the powering of agricultural arms and actuators. When increasing the dynamic efforts and torques that have to be generated, the power systems chosen range from pneumatic, to electric and then hydraulic. The first ones are the easiest to design and implement, because of their much wider use in industrial robotics and automation. Agricultural and forest robots need often to generate high power, and the use of hydraulics is obligatory (Bonicelli *et al.*, 1989b; Brubaker *et al.*, 1990).

Hydraulic system control for robotic applications generally implies a certain level of system modelling in order for the control unit to predict the responses of such fluidic phenomena to any control order, in any situation of the actuator (Ayers *et al.*, 1989; Marchant and Frost, 1989). Complex system representation, such as Bond diagrams, is very useful in these applications (League and Cundiff, 1988; Herman *et al.*, 1990).

Multi-sensing mountings on robots

Use of sensors

The sensing devices that provide data to the robot control system are still the subject of much research activity. However, the use of sensors in agriculture is not simply limited to robots. They can be found in many other applications, including supervision, regulation, forecasting, product evaluation, animal identification, and operator assistance. This wide range of application means that sensors are becoming more and more familiar to both industry and agriculture, and will help to open the way to growing robotics (Werkhoven *et al.*, 1990).

Information delivered by sensors

Climate plays a very important role in agriculture. Data related to temperature, atmospheric pressure, hygrometry and wind speed have therefore to be handled. Very common types of sensors are used for this purpose, and no special mention will be made about them. This section will concentrate instead on those sensors used on agricultural mobile machinery.

Two kinds of information may be delivered by sensors: *internal* and *external* information.

Internal information may be:

- force and weight (Auerhammer *et al.*, 1988);
- torque;
- power;
- consumption;
- skidding or wheel-slip (Chancellor and Zhang, 1988);
- ground speed;
- verticality;
- the angular rotation, acceleration or speed of the machine's frame or chassis (Rabatel and Donicelli, 1986; Zhang and Chancellor, 1989).

Speed regulators on combine harvesters are an example of sophisticated internal sensors; productivity optimization is achieved by checking that the rate of grain loss does not exceed an acceptable maximum. This is achieved by measuring the number of lost grains as they fall against a sheet of metal. The sound produced is electronically analysed and the number of grains hitting the sheet is estimated. This measure is then used in a closed loop control to regulate the speed of the combine.

External sensors deliver a wide variety of information (Byler *et al.*, 1988; Jia and Krutz, 1990; Stafford and Ambler, 1990). For instance, fruit harvesting requires information about several external factors (Sittichareonchai and Sevila, 1989; Bourely *et al.*, 1990; Figs 5.4, 5.5). These include:

- trees and tree row morphology;
- the presence and position of fruits;
- useful parameters on the fruit itself.

External contact with surrounding objects is generally sufficient to sense such information. However, we suggest that it will soon be possible for the internal quality of these objects to be sensed to extend the number of these parameters. Most of the corresponding exploratory work in this area is concerned with on-line inspection of the internal quality of products using sound transmission or Nuclear Magnetic Resonance (Chen *et al.*, 1989; Armstrong *et al.*, 1990; Nelson *et al.*, 1990; Cho *et al.*, 1990; Whalley, 1990).

Different types of sensors

Each automated function requries the use of specific sensors as a source of information. The most basic sensors are tactile ones, which detect a direct mechanical contact with the object and include strain-gauge mountings and piezo-electric devices. Such sensors are used for vehicle guidance or for object detection (Stafford and Hendrick, 1988; Pang and Zoerb, 1990).

A second class of sensors involves non-contact devices using acoustic or electromagnetic waves as a vehicle of information, or reaction to the magnetic or gravitational fields. In some cases objects to be detected can deliver their own signals: artificial objects (beacons) emit acoustic or electromagnetic waves; natural objects emit sounds or infrared waves.

This non-contact transfer allows the transmission of information over a distance revealing presence, distance, nature and/or structure of the object, depending on the task to be performed.

Agricultural transducers may use different kinds of waves. Acoustic (sonic or ultrasonic), as well as electromagnetic waves (gamma and X-rays, and ultra-violet, visible, infra-red and radio emissions) can be used. It should be noted that both types of waves have their drawbacks: acoustic waves are sensitive to atmospheric conditions, such as temperature, humidity and wind, and they produce low directivity; and the high velocity of electromagnetic waves has made it difficult to obtain an accurate spatial resolution with low-cost apparatus.

The simulation of human perceptions and decisions

Simulations of human perceptions and decisions on complex objects and tasks have been necessary ever since the very first automated system was developed. For real-time and on-machine use, such simulations can not be as complex and powerful as those possible in other areas of artificial intelligence application.

On-board computers carried by agricultural machinery currently handle expert systems software of a low level of complexity, but this will change rapidly. The on-board simulation of human expertise will become more and more common (Alvarez *et al.*, 1990). Nowadays, tractors and combines have internal diagnostic sofware giving the driver advice and warnings, using internally sensed information. They will soon be taking an ever-increasing number of global machine-operating decisions, using information from external sensors.

For these operating decision-making tasks, the combination of basic and previously learned know-how and information from sensors is still a novel field of research activity in the field of agricultural robotics. This is particularly true when the sensors are complex (as in image feature extraction, for example: Grand d'Esnon *et al.*, 1988, 1989; Bourely *et al.*, 1990; Marchant and Schofield, 1990), and when their output has to be combined to obtain an appropriate database for decision-making: sensor fusion and neural network techniques are being evaluated for these purposes.

The expansion of agricultural automation

To demonstrate how automation is rapidly expanding in agriculture, following closely the age of mechanization, some examples of the many existing applications of automation will now be described. Crop production involves a greater diversity of automated applications than animal production (Table 5.1).

Table 5.1. Automation applications in crop and animal production.

| Crop production | | Animal production applications |
Application area	Application	
Environment	Environment control	Ethology
Tractors	Tractor control	Environment control
Customized work	Prescribed tillage	Controlled feed
	Water management	Formulation systems
	Planting distribution	Milking sheds
	Pest management	Sheep shearing
	Fertilizer application	Animal identification
	Selective harvesting	
Biotechnology	Bio-engineering	
	Tissue culture	
Product processing	Quality sorting	
	Fruit maturation sensor	
	Robotized packaging	

Examples of automation

Crop production

Irrigation: Sensors provide information on soil moisture (Balascio and Lomas, 1989; Freeland, 1989) and organic soil matter (Pitts, 1986) for customized watering and fertilizing. Automatic control of irrigation is therefore possible (Phene *et al.*, 1986; Werkhoven *et al.*, 1990).

Cultivation: Sensing soil characteristics provides information for the dynamic adjustment of cultivators for seedbed preparation (Stafford and Ambler, 1990).

Chemical spraying; fertilizer application: Automated sprayer attitude or position controls ensure regular spray deposition (Marchant, 1985; Wood and Benneweis, 1985; Monod and Bonicelli, 1986; Giles *et al.*, 1989). Sensors and machine vision can be used to control the efficiency of application (Kranzler *et al.*, 1985; Delwiche *et al.*, 1988; Grenier and Launay, 1989). This means that chemicals are sprayed more precisely and with lower loss, especially through the drift of fine droplets in the wind (Legg and Miller, 1990; Walklate, 1991). A new area of research and commercial development deals with adaptive chemical and fertilizer application according to the position in the field and to the characteristics of the soil–crop system (Ollila *et al.*, 1990; Walsh *et al.*, 1990).

Winter pruning: Trees and vines are usually selectively pruned in winter. Attempts to robotize this complex action have been made, especially on the grapevine (Sevila, 1985; Peyran *et al.*, 1986; Naugle *et al.*, 1989; Sevila *et al.*, 1990).

Tractors and combines: The guidance of tractors was probably the first attempt to introduce electronics in agriculture. As early as the 1930s, it was proved that radio-guided tractors could work (Riddle, 1986). Sensors measure such tractor parameters as ground speed (Baylou *et al.*, 1986), and torque, force and power (Karl-Heinz and Van der Beek, 1986; Auerhammer *et al.*, 1988). The information from the sensors is used to optimize the tractor's performance (Bergmann and Kipp, 1986). A tractor can be also automatically guided while working, in a similar manner to a robotic ploughing system (Bonicelli and Monod, 1987: Stafford and Hendrick, 1988; Brown *et al.*, 1990; Palmer, 1990). The high price of combines requires optimal use by means of automatic control (Jarvenpaa, 1988). Sensors detect grain rates harvested or lost (Baerdemaeker *et al.*, 1985; Pang and Zoerb, 1990), and allow automatic control of cleaning, advance rates and collection of yield data on each field. The combine's grain-unloading tube can be automatically steered (Peltola and Laine, 1990).

Harvesting: Harvesting of soft fuits and vegetables was very rapidly considered suitable for robotization because of its importance and difficulty (Kawamura *et al.*, 1987). Research has been undertaken on various types of harvest, including citrus fruits (Bourely *et al.*, 1990; Juste and Fomes, 1990; Harrell, 1990a,b); apples (Grande d'Esnon *et al.*, 1987; Bourely *et al.*, 1990; Fig. 5.5); asparagus (Ouali, 1986; Humburg and Reid, 1990); grapes (Sittichareonchai and Sevila, 1989; Bourely *et al.*, 1990); flowers (Knani, 1984); and melons (Wolf *et al.*, 1990).

Most harvesters for the above fruits are equipped with visual sensors to detect, locate and estimate the degree of maturity of the produce.

Automatic guidance of lawnmowers has also been developed (Bastard, 1986; Taniwaki *et al.*, 1990).

Greenhouses: An inventory and an inspection can be made of growing plants in nursery beds by image processing (Devoe and Kranzler, 1985; Bennedsen *et al.*, 1990). Seedlings and the supporting benches can be handled automatically (Launay and Grenier, 1988).

Transplanting and tissue culture: Visual sensors have been used for seedling selection (Maw and Suggs, 1984; Tohmaz and Hassan, 1990), and transplanting robots have been developed (Deleplanque *et al.*, 1985; Hwang and Sistler, 1986; Kutz *et al.*, 1987; Brewer, 1990; Chen and Holmes, 1990; Ferrand *et al.*, 1990; Ling *et al.*, 1990; Simonton, 1990a; Ting *et al.*, 1990). Plantlets are inspected by machine vision and appropriately cut for tissue culture (Smith, 1989; Tillett *et al.*, 1990). Other biotechnology approaches include the automation of embryo production (Grand d'Esnon *et al.*, 1988, 1989).

Animal production

Poultry: Egg candling tasks can be robotized (Bourely *et al.*, 1986), and the environment of poultry houses automatically controlled (Mitchell, 1986).

Livestock: Numerous and important research studies are being made into automatic milking (Montalescot, 1986; Bergerot *et al.*, 1989; Frost, 1990; Mechineau *et al.*, 1990 and Fig. 5.6; Schillingmann and Artmann, 1990). Various commercial milking robots will soon be on the market. They will include sensors for detecting illness and oestrus (Wheeler and Graham, 1986a; Onyango *et al.*, 1988; Rossing *et al.*, 1988). Animal supervision is easier with the aid of identification systems and remote sensing devices (Sigrimis *et al.*, 1985; Wheeler and Graham, 1986b; Geers *et al.*, 1988; Marchant and Schofield, 1990). Two robotized sheep shearing systems have been developed in Australia (Key, 1985).

Aquaculture: This new area of farming will be among the first users of automation (Heyerdahl, 1989).

Product control and handling

Sorting with on-line machine vision has been successfully implemented, especially on fruits and vegetables (Bourely *et al.*, 1986; Davenel *et al.*, 1988b; Rigney and Kranzier, 1988; Batchelor and Searcy, 1989; Miller and Delwiche,

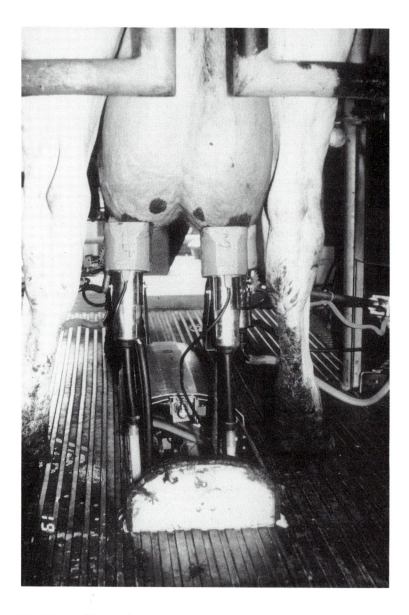

Fig. 5.6. A milking robot for cows. (CEMAGREF, Antony, France.) As the cow comes into position the four robot arms are retracted. The cow is identified, and a machine vision system (see Fig. 5.1) scans the teat positions to compare them with the positions at the previous milking operation. Finally, the arms gently and rapidly place the suction cups on the teats.

1989; Rekhugler and Throop, 1989; Teneze *et al.*, 1989; Throop *et al.*, 1989; Delwiche *et al.*, 1990; Ding *et al.*, 1990; Howarth *et al.*, 1990; Marchant *et al.*, 1990; Tao *et al.*, 1990). Physical sensing on the flesh of the product is also in use (Crochon, 1988; Davenel *et al.*, 1988a; Feller and Margolin, 1988; Zaltzman *et al.*, 1988; Delwiche *et al.*, 1989; Armstrong *et al.*, 1990; Cho *et al.*, 1990).

Conclusion

Technical progress, for a long time concerned with increasing farm machinery power and economic efficiency, now concentrates on developing machines that incorporate a capacity to adapt and select − a capacity restricted, up to now, to the human worker. Computer-aided farming, production simulation, forecasting instruments, and various on-farm expert systems are supposed to help the farm in its evolution. Robotic equipment, especially of the mobile variety, appears to be another key technology for the farms of the year 2000.

Throughout the history of agriculture, humans discovered only one way to prosper: the farmer had to exploit animals, other people (slaves, serfs) and sometimes his own family. Tomorrow he or she will have a more pleasant and peaceful way of prospering, by exploiting the robots working throughout the farm. The farmer's main task, one day, will probably be to control a fleet of them.

References

Alvarez, I., Gauthier, B. and Fuhs, T. (1990) Intelligent supervisor for the tractor−plough system. *Proceedings of the AG-ENG 90, Conference, Berlin, October 1990.* VDI-AGR p1139 D 4000 Dusseldorf, 195−6.

Armstrong, P., Zapp, H. R. and Brown, G. K. (1990) Impulsive excitation of acoustic vibrations in apples for firmness determination. *Transactions of the ASAE* 33(4), 1353−9.

Auerhammer, H., Stanzel, H and Demmel, M. (1988) Electronical weighing equipment for the three-points linkage of tractors. *Proceedings of the AG-ENG 88 Conference, Paris, 1988.* CEMAGREF, Antony 92160, France. AG-ENG paper no. 88.270.

Ayers, P. D., Varma, K. V. and Karim, M. N. (1989) Design and analysis of electro-hydraulic draft control system. *Transactions of the ASAE* 32(6), 1853−5.

Baerdemaeker, J. D., Delcroix, R. and Lindemans, P. (1985) Monitoring the grain flow on combines. *Proceedings of the Agrimation I conference*, ASAE, St Joseph, MI, USA, 329−45.

Balascio, C. C. and Lomas, K. M. (1989) A comparative study of moisture sensors for use in mushroom beds. *Transactions of the ASAE* 32(3), 928–33.

Bastard, D. (1986) Projet de robot mobile pour application à la tonte du gazon. *Proceedings of the Agrotique 86 Conference, Bordeaux, France*, ADESO, 83–99.

Batchelor, M. M. and Searcy, S. W. (1989) Computer vision determination of the stem/root joint on processing carrots. *Journal of Agricultural Engineering Research* 43, 259–69.

Baylou, P., Bousseau, G. and Thevenin, A. (1986) Etude et réalisation d'un capteur à ultrasons mesurant la vitesse d'un tracteur agricole. *Proceedings of the Agrotique 86 Conference, Bordeaux, France*. ADESO, 111–17.

Bennedsen, B. S., Feuilloley, P. and Grand d'Esnon, A. (1990) Colour vision for inspection of pot plants. *Proceedings of the AG-ENG 90 Conference, Berlin, October 1990*. VDI-AGR p1139, D.4000 Dusseldorf, Germany, 345–6.

Bergerot, P., Baylou, P. and Ordolff, D. (1989) Système de pose automatique de gobelets trayeurs. *Agrotique 89: Proceedings of the Second International Conferance, Bordeaux*. TEKNEA, 13241 Marseille, France, 317–30.

Bergmann, E., and Kipp, J. C. (1986) Experiences on optimisation algorithms for heavy tractor operations. *Proceedings of the Agrimation II Conference*. ASAE, St Joseph, MI, USA.

Bonicelli, B., and Monod, M. O. (1987). A self-propelled plowing robot. ASAE paper No. 87.1064. ASAE, St Joseph, MI, USA.

Bonicelli, B., Lucas, J. and Perret, F. (1989a) Utilisation des techniques robotiques en mécanisation forestière. *Agrotique 89: Proceedings of the Second International Conference, Bordeaux*. TEKNEA, 13241 Marseille, France, 295–306.

Bonicelli, B., Lucas, L., Perret, F. and Bonnafous, J. C. (1989b) Robot arm for forest thinning RAFT. ASAE paper No. 89 7056. ASAE, St Joseph, MI, USA.

Bourely, A., Hsia, T.C. and Upadhyaya, S. K., (1986) Investigation of a robotic egg candling system. *Proceedings of the Agrimation II conference*, ASAE, St Joseph, MI, USA, 53–62.

Bourely, A. (1989) Un contrôleur polyvalent pour robots agro-industriels. *Agrotique 89: Proceedings of the Second International Conference, Bordeaux*. TEKNEA, 13241 Marseille, France, 387–93.

Bourely, A., Rabatel, G., Grand d'Esnon, A. and Sevila, F. (1990) Fruit harvest robotization: 10 years of CEMAGREF experience on apple, grape and orange. *Proceedings of the AG-ENG 90 Conference, Berlin, October 1990*. VDI-AGR p1139, D.4000 Dusseldorf, Germany, 178–9.

Brewer, H. L. (1990) Automated devices to grow and transplant seedlings. *IROS'90. IEEE – Proceedings, vol 1: Conference*. Tsuchiura, Ibaraki, Japan, July 1990. IEEE, NJ 08854, USA, 243–8.

Brown, H. N., Wilson, J. N. and Wood, H. C. (1990) Image analysis for vision-based agricultural vehicle guidance. ASAE paper No. 90.1623. ASAE, St Joseph, MI, USA.

Brubaker, J. R., Krutz, G. W., Nine, P. L. and Ukrainetz, P. R. (1990) Detasseler electrohydraulic position control. ASAE paper No. 90.1634. ASAE, St Joseph, MI, USA.

Buschmeier, R. (1990) Computer-aided farming with the satellite navigation system GPS. *Proceedings of the AG-ENG 90 Conference, Berlin, October 1990*. VDI-AGR p1139, D.4000 Dusseldorf, Germany, 88–9.

Byler, R. K., Kromer, K. H. and Koslav, M. (1988) Sensor technology for soil surface measurement. *Proceedings of the AG-ENG 88 Conference, Paris 1988.* CEMAGREF, Antony 92160, France. AG-ENG paper No 88.264.

Chancellor, W. and Zhang, N. (1989) Automatic wheel-slip control for tractor. *Transactions of the ASAE* 32(1), 17–22.

Chen, C. and Holmes, R. G. (1990) Simulation and optimal design of a robot arm for greenhouse/nursery. ASAE paper No 90.1537. ASAE, St Joseph, MI, USA.

Chen, P., McCarthy, M. J. and Kauten, R. (1989) NMR for internal quality evaluation of fruits and vegetables. *Transactions of the ASAE* 32(5), 1747–53.

Cho, S. I., Krutz, G. W. and Bellon, V. (1990) Non-destructive ripeness sensing by using proton NMR. *Proceedings of the AG-ENG 90 Conference, Berlin, October 1990.* VDI-AGR p1139, D.4000 Dusseldorf, Germany.

Choi, C. H., Erbach, D. C. and Smith, R. J. (1990 Navigational tractor guidance system. *Transactions of the ASAE* 33(3), 699–706.

Crochon, M. (1988) A microrefractometer for fruit internal quality in site measurement. *Proceedings of the AG-ENG 88 Conference, Paris, 1988.* CEMAGREF, Antony 92160, France. AG-ENG paper No. 88.310.

Davenel, A., Sevila, F., Crochon, M., Pourcin, J., Verlaque, P., Bertrand, D. and Robert, P. (1988a) On line infra-red measurement in food processing. *Proceedings of the AG-ENG 88 Conference, Paris, 1988.* CEMAGREF, Antony 92160, France. AG-ENG paper No. 88.308.

Davenel, A., Guizard, C., Labarre, T. and Sevila, F. (1988b) Automatic detection of surface defects on fruits by using a vision system. *Journal of Agricultural Engineering Research* 41, 1–9.

Defranco, H. (1988) Device for self-steering by checking a crop's row. *Proceedings of the AG-ENG 88 Conference, Paris, 1988.* CEMAGREF, Antony 92160, France. AG-ENG paper No. 88.388.

Deleplanque, H., Bonnet, P. and Postaire, J. G. (1985) An intelligent robotic system for in vitro plantlet production. *Proceedings of RoViSec 5.* Fifth International conference on Robot Vision and Sensory Control, 29–31 October 1985. Amsterdam, The Netherlands.

Delwiche, M. J., Tang, S and Mehlschau, J. J. (1988) A fruit firmness sorting system. *Proceedings of the AG-ENG 88 Conference, Paris, 1988.* CEMAGREF, Antony 92160, France. AG-ENG paper No. 88.230.

Delwiche, M. J., Tang, S. and Mehlschau, J. J. (1989). An impact force response fruit firmness sorter. *Transactions of the ASAE* 32(1), 321–6.

Delwiche, M. J., Tang, S. and Thompson J. F. (1990) Prune defect detection by line-scan imaging. *Transactions of the ASAE* 33(3), 950–4.

DeVoe, D. R., and Kranzler, G. A. (1985) Image processing to inventory seedlings in nursery beds. *Proceedings Agrimation I Conference.* ASAE, St Joseph, MI, USA, 215–22.

Ding, K., Morey, R. V., Wilcke, W. F., Hansen, D. J. (1990). Corn quality evaluation with computer vision. ASAE paper No. 90.3532, ASAE, St. Joseph, MI, USA.

Feller, R. and Margolin, E. (1988) Separation of clods and stones on a potato harvester. *Proceedings of the AG-ENG 88 Conference, Paris, 1988.* CEMAGREF, Antony 92160, France. AG-ENG paper No. 88.252.

Ferrand, C., Baylou, P. and Grenier, G. (1990) A non-damaging effector for high speed transplanting robots. ASAE paper No. 90.7501. ASAE, St Joseph, MI, USA.

Freeland, R. S. (1989) Review of soil moisture sensing using soil electrical conductivity. *Transactions of the ASAE* 32(6), 2190–3.

Frost, A. R. (1990) Robotic milking: A review. *Robotica* 8, 311–18.

Geers, R., Goedseels V. and Berkmans, D. (1988) The evaluation of skin temperatures as a sensory system for robotic control of animal husbandry processes. *Proceedings of the AG-ENG 88 Conference, Paris, 1988*. CEMAGREF, Antony 92160, France. AG-ENG paper No. 88.064.

Giles, D. K., Delwiche, M. J. and Dodd, R. B. (1988) Electronic measurement of tree canopy volume. *Transactions of the ASAE* 31(1), 264–72.

Giles, D. K., Delwiche, M. J. and Dodd, R. B. (1989) Sprayer control by sensing orchard crop characteristics: Orchard architecture and spray liquid savings. *Journal of Agricultural Engineering Research* 43, 271–89.

Grand d'Esnon, A., Rabatel, G., Pellenc, R., Journeau, A. and Aldon, M. J. (1987) Magali: A self-propelled robot to pick apples. ASAE paper No. 87-1037. ASAE, St Joseph, MI, USA.

Grand d'Esnon, A., Harrell, R. C., and Chee, R. (1988) Evaluation by quality and quantity of tissue culture in liquid medium. *Proceedings of the AG-ENG 88 Conference, Paris, 1988*. CEMAGREF, Antony 92160, France. AG-ENG paper No. 88.395.

Grand d'Esnon, A., Sujian, N., Faure, S. and Sevila, F. (1989) On line evaluation by vision systems in bio-technologies. ASAE paper No. 89.7057. ASAE, St Joseph, MI, USA.

Grenier, G. and Launay, I. (1989) Automatisation des traitements phytosanitaires en serre conception d'un module adaptable sur robot de manutention. *Agrotique 89: Proceedings of the Second International Conference, Bordeaux*. TEKNEA, 13241 Marseille, France, 195–206.

Guul-Somonsen, F (1990) Requirements to electronics in agriculture. *Proceeding of the AG-ENG 90 Conference, Berlin, October 1990*. VDI-AGR p1139, D.4000 Dusseldorf, Germany, 354–5.

Han, Y. J. and Hayes, J. C. (1990) Soil cover determination by image analysis of textural information. *Transactions of the ASAE* 33(2), 681–6.

Harrell, R. C., Adsit, P. D., Munilla, R. D. and Shaughter, D. C. (1990a) Robotic picking of citrus. *Robotica* 8, 269–78.

Harrell, R. C., Adsit, P. D., Pool, T. A. and Hoffman, R. (1990b) The Florida robotic grove-lab. *Transactions of the ASAE* 33(2), 391–9.

Hermann, T., Bonicelli, B. and Mansion, M. (1990) Bond-graph and dynamic modeling of a high power forestry-arm. *Proceedings of the AG-ENG 90 Conference, Berlin, October 1990*. VDI-AGR p1139, D.4000 Dusseldorf, Germany, 365–6.

Heyerdahl, P. H. (1989) Onshore feeding robot in aquaculture. *Agrotique 89: Proceedings of the Second International conference, Bordeaux*. TEKNEA 13241 Marseille, France, 207–19.

Howarth, M. S., Brandon, J. R., Searcy, S. W. and Kehtarnavaz, N. (1990) Estimation of tip shape for carrot classification by machine vision. ASAE paper No. 90.3530. ASAE, St Joseph, MI, USA.

Humburg, D. S. and Reid, J. F. (1990) Field performance of machine vision for the selective harvest of asparagus. ASAE paper No. 90.7523. ASAE, St Joseph, MI, USA.

Hwang, H. and Sistler, F. E. (1986) A robotic pepper transplanter. *Applied Engineering in Agriculture* 2(1), 2–5.

Ito, N. (1990) Agricultural robots in Japan. *IROS 90. IEEE, Proceedings vol. 1; Conference.* Tsuchiura, Ibaraki, Japan, July 1990. IEEE, NJ 08854 USA, 249–54.

Jarvenpaa, M. (1988) The advantages of improved information systems and automatic control of a combine harvester. *Proceedings of the AG-ENG 88 Conference, Paris, 1988.* CEMAGREF, Antony 92160, France. AG-ENG paper No. 88.167.

Jia, J. and Krutz, G. W. (1990) Identification and location of the maize plant with machine vision. *Proceedings of the AG-ENG 90 Conference, Berlin, October, 1990.* VDI-AGR p1139, D.4000 Dusseldorf, Germany, 86–7.

Juste, F. and Fornes, I. (1990) Contributions to robotic harvesting in citrus in Spain. *Proceedings of the AG-ENG 90 Conference, Berlin, October 1990.* VDI-AGR p1139, D.4000 Dusseldorf, Germany, 146–7.

Karl-Heinz, M. and Van der Beek, A. (1986) Electronics for smarter tractor/implement systems. *Proceedings of the Agrimation II Conference.* ASAE, St Joseph, MI, USA.

Kawamura, N., Namikawa, K., Fujiura, T., Ura, M. and Ogawa, Y. (1987) Study on agricultural robot VII: Hand of fruit harvesting robot. *Research Report on Agricultural Machinery* 17, 1–7. Kyoto University, Kyoto, Japan.

Key, S. J. (1985) Productivity, modelling and forecasting for automated shearing machinery. *Proceedings of the Agrimation I Conference.* ASAE St Joseph, MI, USA, 200–9.

Knani, G. (1984) 'Système sensoriel (navigation et travail) d'un robot de cueillette de fleurs.' Doctoral thesis, University of Bordeaux, France.

Kranzler, G. A. and others (1985). Computer vision for evaluation of agricultural chemical application. *Proceedings of the Agrimation I Conference.* ASAE, St Joseph, MI, USA, 136–41.

Kutz, L. J., Miles, G. E., Hammer, P. A. and Krutz, G. W. (1987) Robotic transplanting of bedding plants. *Transactions of the ASAE* 30(3), 586–90.

Launay, Y. and Grenier, G. (1988) A new conception of automated production in a horticultural greenhouse. *Proceedings of the AG-ENG 88 Conference, Paris, 1988.* CEMAGREF, Antony 92160, France. AG-ENG paper No. 88.234.

League, R. B. and Cundiff, J. S. (1988) Bond graph model of a hydrostatic drive test stand. *Transactions of the ASAE* 31(1), 28–36.

Legg, B. J. and Miller, P. C. (1990) Drift assessment using measurements and mathematical models. ASAE paper No. 90.1593. ASAE, St Joseph, MI, USA.

Ling, P. P., Tai, Y. W. and Ting, K. C. (1990) Vision guided robotic seedling transplanting. ASAE paper 90-7520. ASAE, St Joseph, MI, USA

Lucas, J. (1984) A propos de la robotisation de l'agriculture. *Bulletin technique du machinisme et de l'équipement* 4 CEMAGREF, Antony 92160, France.

McDonald, Y. and Chen, Y. R. (1990) Application of morphological image processing in agriculture. *Transactions of the ASAE* 33(4), 1345–52.

Marchant, J. A. (1985) The use of robotics in the agricultural and food industries. Div. Note DN 1304, National Institute for Agricultural Engineering, Silsoe, UK, November.

Marchant, J. A. and Frost, A. R. (1985) Spray boom attitude control system. *Proceedings of the Agrimation I Conference.* ASAE, St Joseph, MI, USA, 142–7.

Marchant, J. A. and Frost, A. R. (1989) Simulation of the performance of state feedback controllers for an active spray boom suspension. *Journal of Agricultural Engineering Research* 43, 77–91.

Marchant, J. A. and Schofield, C. P. (1990) Image analysis for live animals: Some techniques and results. *Proceedings of the AG-ENG 90 Conference, Berlin, October 1990.* VDI-AGR p1139, D.4000 Dusseldorf, Germany, 68–9.

Marchant, J. A., Onyango, C. M. and Elipe, E. (1989) Weight and dimensional measurements on potatoes at high speed using image analysis. *Agrotique 89: Proceedings of the Second International Conference, Bordeaux.* TEKNEA, 13241 Marseille, France, 41–52.

Marchant, J. A., Onyango, C. M. and Street, M. J. (1990) Computer vision for potato inspection without singulation. *Computers and Electronics in Agriculture* 4, 235–44.

Maw, B. W. and Suggs, C. W. (1984) Sorting and selection of bare roots transplants. *Transactions of the ASAE* 27(3).

Mechineau, D. (1988) Absolute field location. *Proceedings of the AG-ENG 88 Conference, Paris, 1988.* CEMAGREF, Antony 92160, France. AG-ENG paper No. 88.393.

Mechineau, D., Montalescot, J. B., Bonneau, D., Vigneau, J. L., Marchal, P., Rault, G. and Collewet, C. (1990) Milking robot. ASAE paper 90-7047. ASAE, St Joseph, MI, USA.

Miller, B. K. and Delwiche, M. J. (1989) A color vision system for peach grading. *Transactions of the ASAE* 32(4), 1484–90.

Mitchell, B. W. (1986) Microcomputer based environmental control system for a disease free poultry house. *Transactions of the ASAE* 29(4).

Molto, E., Juste, F. and Pla, F. (1990) Detection of citrus fruits by vision systems in robotic harvesting. *Proceedings of the AG-ENG 90 Conference, Berlin, October 1990.* VDI-AGR p1139, D.4000 Dusseldorf, Germany, 337–8.

Monod, M. O. and Bonicelli, B. (1986) Stabilisation d'une rampe de pulvérisation. *Proceedings of the Agrotique 86 Conference, Bordeaux, France.* ADESO, 143–54.

Monod, M. O and Mechineau, D. (1986) Localisation absolue et application potentielle à un matèriel automoteur. *Proceedings of the Agrotique 86 conference, Bordeaux, France.* ADESO, 119–31.

Monsion, M. (1985). Visual servo-control in automatic winter pruning of the grapevine. *IEE International Conference: Control 85.* Cambridge, MA. IEEE, NJ 08854, USA.

Montalescot, J. B. (1986) L'électronique et les automatismes en élevage laiter: vers la robotisation de la traite. *Proceedings of the Agrotique 86 conference, Bordeaux, France.* ADESO, 243–63.

Naugle, J. A., Rehkugler, G. E. and Throop, J. A. (1989) Grapevine cordon following using digital image processing. *Transactions of the ASAE* 32(1), 309–15.

Nelson, S. O., Lawrence, K. C., Kandala, C. V. K., Himmelsbach, D. S., Windham, W. R. and Kraszewskio, A. W. (1990) Comparison of DC conductance, RF impedance, microwave, and NMR methods for single-kernel moisture measurement in corn. *Transactions of the ASAE* 33(3), 893–8.

Ollila, D. G., Schumacher, J. A. and Froehlich, D. P. (1990) Integrating field grid sense system with direct injection technology. ASAE paper No. 90.1628. ASAE, St Joseph, MI, USA.

Onyango, C. M., Marchant, J. A., Lake, J. R. and Stanbridge, D. A. (1988) A low maintenance conductivity sensor for detecting mastitis. *Journal of Agricultural Engineering Research* 40, 215–24.

Ouali, K. (1986) 'Synthèse des fonctions navigation et cueillette d'un robot agricole ramasseur d'asperges.' Doctoral thesis, University of Bordeaux.

Palmer, R. (1990) Positioning requirements for in-field spatial variability, ASAE paper No. 90.1622. ASAE, St Joseph, MI, USA.

Pang, S. N. and Zoerb, G. C. (1990) A grain flow sensor for yield mapping. ASAE paper No. 90.1633. ASAE, St Joseph, MI, USA.

Peltola, A. and Laine, E. (1990) Automatic steering system for forage harvester's blowing tube. *Proceedings of the AG-ENG 90 Conference, Berlin, October 1990.* VDI-AGR p1139, D.4000 Dusseldorf, Germany, 140–1.

Peyran, B., Monsion, M., Baylon, P. and Genet, P. (1986) Description d'un robot autonome destiné à la taille automatique de la vigne. *Proceedings of the Agrotique 86 Conference, Bordeaux, France.* ADESO, 59–69.

Phene, C. J. and others (1986) Automated feedback irrigation scheduling and control with a weighted lysimeter. *Proceedings of the Agrimation II conference.* ASAE, St Joseph, MI, USA.

Pitts, M. J., Hummel, J. W. and Butler, B. J. (1986) Sensors utilizing light reflection to measure soil organic matter. *Transactions of the ASAE* 29(2).

Rabatel, G. (1988) A vision system for the fruit picking robot Magali. *Proceedings of the AG-ENG 88 Conference, Paris, 1988.* CEMAGREF, Antony 92160, France. AG-ENG paper No. 88.293.

Rabatel, G. and Bonicelli, B. (1986). Stabilisation d'assiette et d'horizontalité latérale sur engin forestier. *Proceedings of the Agrotique 86 Conference, Bordeaux, France.* ADESO, 133–54.

Rehkugler, G. E. and Throop, J. A. (1986) Apple sorting with machine vision. *Transactions of the ASAE* 29(5), 1388–97.

Rehkugler, G. E. and Throop, J. A. (1989) Image processing algorithm apple defect detection. *Transactions of the ASAE* 32(1), 267–72.

Reid, J. F. and Searcy, S. W. (1988) An algorithm for separating guidance information from row crop images. *Transactions of the ASAE* 31(6), 1624–32.

Riddle, W. E. (1986) A time of technological change. Agriculture 2000. *Proceedings of the Agrimation I Conference.* ASAE, St Joseph, MI, USA, 21–6.

Rigney, M. P. and Kranzler, G. A. (1988) Machine vision for grading southern pine seedlings. *Transactions of the ASAE* 31(2), 642–6.

Rossing, R., Benders, E., Hogewerf, P. H., Hopster, H. and Maatje, K. (1988) Real-time measurements of milk conductivity for detecting mastitis. *Proceedings of the AG-ENG 88 Conference, Paris, 1988.* CEMAGREF, Antony 92160, France. AG-ENG paper No. 88.376.

Schillingmann, D. and Artmann, R. (1990) Robot milking–Development of a robot system and first experiments. *Proceedings of the AG-ENG 90 Conference, Berlin, October 1990.* VDI-AGR p1139, D.4000 Dusseldorf, Germany, 66–7.

Sevila, F. (1985) A robot to prune the grapevine. *Proceedings of the Agrimation I Conference.* ASAE, St Joseph, MI, USA, 190–9.

Sevila, F., Piquet, J. C. and Baylou, P. (1990) Computer vision system and growth modelization software applied to grapevine woods and arms, as pruning robot design tools. *Proceedings of the AG-ENG 90 Conference, Berlin, October 1990.* VDI-AGR p1139, D.4000 Dusseldorf, Germany, 376–7.

Shmulevich, I., Zeltzer, G. and Brunfield, A. (1987) Guidance system for field machinery using laser scanning method. ASAE paper No. 87.1558. ASAE, St Joseph, MI, USA.

Shmulevich, I., Zeltzer, G. and Brunfield, A. (1989) Laser scanning method for guidance of field machinery. *Transactions of the ASAE* 32(2), 425–30.

Sigrimis, N. A., Scott, N. R. and Csarnieckly, C. S. (1985) A passive transponder identification system for livestock. *Transactions of the ASAE* 28(2).

Simonton, W. (1990a) Robotic workcell for greenhouse propagation. *Proceedings of the AG-ENG 90 Conference, Berlin, October 1990.* VDI-AGR p1139, D.4000 Dusseldorf, Germany, 347–8.

Simonton, W. (1990b) Robotic end effector for handling greenhouse plant material. ASAE paper No. 90.7503. ASAE, St Joseph, MI, USA.

Sittichareonchai, A. and Sevila, F (1989) A robot to harvest grapes. ASAE Paper No. 89-7074. ASAE, St Joseph, MI, USA.

Slaughter, D. C. and Harrell, R. C. (1989) Discriminating fruit for robotic harvest using color in natural outdoor scenes. *Transactions of the ASAE* 32(2), 757–63.

Smith, G. P. (1989) The application of robotics to horticultural micropropagation. *Agrotique 89: Proceedings of the Second International Conference, Bordeaux.* TEKNEA, 13241 Marseille, France, 179–94.

Stafford, J. V. and Hendrick, J. G. (1988) Dynamic sensing of soil pans. *Transactions of the ASAE* 31(1), 9–13.

Stafford, J. V. and Ambler, B. (1990) Sensing spatial variability of seedbed structure. ASAE paper No. 90.1624. ASAE, St Joseph, MI, USA.

Taniwaki, K., Kamiyama, H. and Soga, T. (1990) Automated steering control of robotic mower. IROS'90, IEEE: *Proceedings, vol 1: Conference.* Tsuchiura, Ibaraki, Japan, July 1990. IEEE, NJ 08854, USA, 280–6.

Tao, Y., Morrow, C. T., Heinemann, P. H. and Sommer, J. H. (1990) Automated machine vision inspection of potatoes. ASAE paper No. 90.3531. ASAE, St Joseph, MI, USA.

Teneze, B., Baylou, P. and Riboulet, P. (1989) Presentation de la partie vision d'un programme 'tri automatique de haricots verts' élaboré par Agrotec. *Agrotique 89: Proceedings of the Second International Conference, Bordeaux.* TEKNEA, 13241 Marseille, 53–66.

Throop, J. A., Rehkugler, G. E. and Upschurch, B. L. (1989) Application of computer vision for detecting watercore in apples. *Transactions of the ASAE* 32(6), 2087–92.

Tillet, R. D. (1989) A calibration system for vision-guided agricultural robots. *Journal of Agricultural Engineering Research* 42, 267–73.

Tillet, N. D. and Nybrant, T. G. (1990) Leader cable guidance of an experimental field gantry. *Journal of Agriculture Engineering Research* 45, 253–67.

Tillett, R. D., Brown, F. R., McFarlane, N. J. B., Onyango, C. M., Davis, P. F. and Marchant, J. A. (1990) Image-guided robotics for the automation of micropropagation. *IROS'90. IEEE. Proceedings, vol 1: Conference.* Tsuchiura, Ibaraki, Japan, July 1990. IEEE, NJ 08854, USA, 265–70.

Ting, K. C., Giacomelli, G. A., Shen, S. H. and Kabala, W. P. (1990) Robot workcell for transplanting of seedlings. *Transactions of the ASAE* 33(3), 1005–17.

Tohmaz, A. S. and Hassan, A. E. (1990) Use of vision system for recognizing plantable seedlings. ASAE paper No. 90.7518. ASAE, St Joseph, MI, USA.

Tutle, E. G. (1983) Image controlled robotics in agricultural environments. Robotics and intelligent machines in agriculture. ASAE, St Joseph, MI, USA.

Walklate, P. J. (1991) Pesticide drift from air-assisted orchard sprayers – a numerical simulation study. *Proceedings* of symposium: 'Air-assisted spraying in crop protection', organized by the Association of Applied Biologists and the British Crop Protection Council. University College of Swansea, UK, January 1991, 61–8.

Walsh, J. L., Ross, C. C., Campbell, D. P., Hartman, N. F., Boswell, F. C. and Hargrove, W. L. (1990) Integrated optic sensor for cropland ammonia volatilization measurement. ASAE paper No. 90.1632. ASAE, St Joseph, MI, USA.

Werkhoven, C., Bosma, A. H., Gabriels, P. C. J. and Stanghellini, C. (1990) Sensors for irrigation scheduling of cultures in the field. *Proceedings of the AG-ENG 90 Conference, Berlin, October 1990.* VDI-AGR p1139, D.4000 Dusseldorf, Germany, 96–7.

Whalley, W. R. (1990) Development of a microwave soil moisture sensor for incorporation in a narrow cultivator tine. *Proceedings of the AG-ENG 90 Conference, Berlin, October 1990.* VDI-AGR p1139, D.4000 Dusseldorf, Germany, 225–6.

Wheeler, P. A. and Graham, K. L. (1986a) A computerized milk temperature conductivity device for dairy use. *Proceedings of the Agrimation II Conference.* ASAE, St Joseph, MI, USA.

Wheeler, P. A. and Graham, K. L. (1986b) A review of remote sensing techniques of dairy cattle. *Proceedings of the Agrimation II Conference.* ASAE, St Joseph, MI, USA.

Whitfield, R. D. (1988) A calibration system for vision guided agricultural robots. *Proceedings of the AG-ENG 88 Conference, Paris, 1988.* CEMAGREF, Antony 92160, France. AG-ENG paper No. 88.124.

Wolf, I., Bar-Or, J., Edan, Y. and Pepier, U. M. (1990) Developing grippers for a melon harvesting robot. ASAE paper No. 90.7504. ASAE, St Joseph, MI, USA.

Wood, H. C. and Benneweis, R. K. (1985) Microcomputer monitoring and control of field sprayer. *Proceedings Agrimation I conference.* ASAE, St Joseph, MI, USA, 142–7.

Young, R. E., Olson, J. and Jahns, G. (1976) *Automatic Guidance of Farm Vehicles. A Monograph.* Agricultural Engineering Department Series No. 1. Auburn University, Auburn, AL, USA.

Zaltzman, A., Verma, B. P., Blankenship, P. D. and Prussia, S. E. (1988) Separating peanuts products based on their physical properties. *Proceedings of the AG-ENG 88 Conference, Paris, 1988.* CEMAGREF, Antony 92160, France, AG-ENG paper No. 88.101.

Zhang, N. and Chancellor, W. (1989) Automatic ballast position control for tractors. *Transactions of the ASAE* 32(4), 1159–64.

Chapter 6

Image Analysis in Biological Systems

J.A. Marchant

Introduction

Although the precise definition of image analysis may vary from one authority to another, it is generally taken to be the processing of an image by a computer to extract meaningful information automatically. Some authors also use the terms 'image processing', 'picture processing', or 'computer vision', which may have slightly different shades of meaning, but here they will be taken to be synonymous.

An important distinction can, however, be drawn between image analysis and image enhancement. Computers can be used very effectively to convert visual data to a form which makes human interpretation easier. For instance, by removing blurring if the camera or the subject is moving, by filtering to remove noise or to enhance edges, or by transforming the image to increase the contrast. This enhancement of an image leaves the extraction of information to a human and, although it may be a precursor to automatic analysis, does not include the analysis itself.

Practical image analysis systems have been available certainly since the early 1970s and possibly even before that. Early uses include the automatic analysis of cervical smear data and inspection of metals for inclusions or impurities. Because of the state of technological development at that time, processing was either done by large and expensive computers taking a great deal of time or by dedicated hardware operating very simple repetitive algorithms on the data. Typical measurements would be the areas of regions contrasting with the background, the shapes of regions, the number of holes within a region and so on.

The development of image analysis systems has followed that of computing. The availability of microprocessors in the early 1970s meant that computing could be done cheaply and in compact devices with little power consumption. Since the development of very powerful 16 and 32 bit microprocessors in the mid-1980s, along with the availability of dense and cheap memory chips, the number of suppliers of image analysis hardware has mushroomed. It is an interesting fact that each electronic card in these modern devices (and in some cases each individual chip) is more powerful than the roomful of units of the original image analysis systems.

Image analysis hardware

Cameras

The sensing element of an image analysis system is the camera. Cameras can be divided into two categories, tube and solid state; the latter type can be further subdivided into area and linescan cameras.

As their name suggests, tube cameras use a vacuum tube to convert the optical image into an electronic signal. Although there are different types of tubes, by far the most common is the Vidicon. In this device a transparent conducting layer is deposited onto the inside front face of the tube. On top of this deposit is a second layer consisting of many small resistive particles. The resistance of these particles changes with illumination. The rear end of the tube contains a cathode producing a beam of electrons which is scanned over the photosensitive layer. The current carried by the beam is modulated by the resistance of the particles at each point in the scan and hence by the intensity of light falling on the front face of the tube. By measuring the current, a signal is derived which has effectively encoded the optical image focused onto the front face of the tube.

Standards for normal television broadcasts in the UK dictate that the whole field of view is divided into 625 lines and scanned completely 25 times a second − the so called frame rate. In order to reduced the flickering effect caused by too low a rate, the frame is divided into two fields, consisting of the odd numbered and the even numbered scan lines. Interlaced fields are thus scanned and presented to the viewer 50 times each second.

By far the most common type of solid state camera, and the only one that will be considered here, is the charge coupled device (CCD). CCD arrays are accommodated typically on a device about 20 mm square and a few millimetres thick, onto which the image is focused. Their construction resembles MOSFETs (metal oxide semiconductor field effect transistors) and so they can be made by well established mass-production methods used for integrated semiconductor chips. A CCD array consists of tens of thousands of individual MOS elements arranged in a grid. Photons from the focused image generate charges within each element, and the array can be scanned to extract a video signal. Colour images can be accommodated by using three arrays, each fitted with a filter to make it sensitive to one of the primary colours; or when lower quality will suffice, by using a special striped filter on a single array. Although CCD cameras are inherently digital devices their output is normally converted to an analogue form to suit standard television signals.

A special type of solid state device is the linescan camera. This device gives an output corresponding to the light intensity along a single line across the image. This may seem not to be of much use until one realizes that there are many situations where the objects of interest move at a steady rate relative to the camera – for example, where the camera is inside an aeroplane surveying the ground below or when objects are moving along a conveyor belt. Advantages of linescan cameras are that much less memory is needed to store the data and the sensor is (at least intrinsically) cheaper. However, neither of these considerations is very important today. The memory size to hold a full image is by today's standards small. Also, even though the sensor has fewer elements, the more generally useful CCD array is produced in much greater numbers for such applications as surveillance and, more importantly, the large consumer market. CCD cameras are mass-produced and therefore cheap.

Digitization and storage

After encoding into an electronic signal, the image must be stored in a digital form suitable for access by the computer. The first operation is to digitize the image. Fig. 6.1 shows an image of a typical agricultural object; (a) shows the whole image while (b) shows part of the image expanded. It can be seen that the image has been divided into an array of so-called 'pixels'. This is done by converting the analogue voltage corresponding to the brightness of a small part of the image into a digital number and storing the number in memory. Fig. 6.1(b) in fact shows the memory contents converted back into analogue form and displayed on a screen. Intensity values are digitized typically using an eight-bit analogue-to-digital converter (256 intensity values or grey levels). The speed of this device has to be fast enough to do the conversion as the data is produced. With an image consisting of 512 horizontal × 512 vertical pixels refreshed 25 times a second this means digitizing at the rate of over 6½ million bytes per second. In fact the rate is normally faster than this as many systems use only part of the field of view to ensure the pixels are square. This requires a rate of the order of 10 million bytes per second. Although such a rate is difficult even to imagine, fortunately this aspect of image analysis is the preserve of the chip designer rather than the analyst and devices for performing these conversions are commonly available.

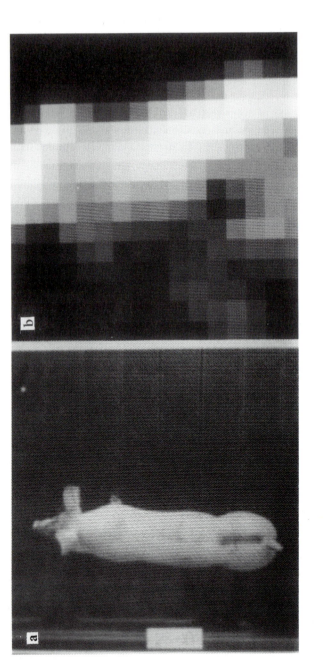

Fig. 6.1. (a) Image of a biological object; (b) part of the image expanded to show digitization.

Although it is possible to do very simple analysis without storing an image, storage makes the process much more easily managed. Usually the whole of an image is stored before analysis starts and a system may have capacity to store several images – for example, to hold intermediate results. To ensure rapid access to data, images must in practice he held in RAM (random access memory). This means that a reasonably sized system will have a RAM capacity for images of, say, 2 Mbyte. As well as being dense, the memory must also be fast; using the figures above the memory must be capable of being written to in less than 100 ns (1 ns = 1 nanosecond = 1 billionth of a second). Once again the design of such devices has already been achieved, and at the time of writing they can be accommodated on a board area about the size of a couple of postage stamps.

Data processing

At the beginning of this chapter, image analysis was defined to be the processing of an image to extract meaningful information. If the image data is changing only slowly, or the subsequent use of this information is infrequent, then a computer of modest power, for example a typical office desktop computer, may well suffice. (Although developing algorithms on a system which takes an hour or so to execute them would not be easy.) On the other hand, there are many situations, such as control of a machine or process, where the image data must be analysed frequently – up to the speed at which it is arriving from the camera, so-called 'real-time' operation. With a standard television system the ultimate might be to analyse 512×512 pixels refreshed 25 times per second using three bytes per pixel, one for each primary colour. This gives the rather daunting rate of nearly 20 million bytes per second. A typical modern microprocessor, say a Motorola 68030 or Intel 80386, may have a clock speed of 25 MHz with a simple instruction taking eight clock cycles. Even if processing required only ten instructions per byte (a very conservative estimate indeed), the processor would still be about 70 times too slow. Also, even though this rate has been termed 'ultimate' above, it it not difficult to find a food production task that approaches it (see the case study later in this chapter).

One way of dealing with the problem of processing speed is not to use a computer at all, but instead to use dedicated electronic circuits to suit a particular task. For example, it is relatively easy to design a circuit which will measure the number of pixels above a certain grey level in an image in real time. If all our objects were high contrast and the processing task was to measure area, this simple solution would suffice. If, however, the situation changed slightly so that we could not guarantee consistent contrast, or we had to measure shape as well as area, the circuitry would need to be redesigned. There would be little flexibility in the system. As will be seen in the next section it is important to retain flexibility to cope with the variable nature of biological objects.

A second course of action is to use dedicated subsystems for commonly needed image processing functions such as edge extraction or boundary following. These can then be called upon by the microprocessor when required. This is the basis of the MaxVideo system developed by Datacube where special purpose electronic cards can be linked to a general purpose image store and processing unit. Also single chip devices are now becoming available for incorporation by system designers; examples include the Plessey PDSP16488 convolver chip and the LSI Logic L64290 contour tracer.

Parallel processing is an attractive option in image analysis. Often an image can be broken down into smaller regions and analysis can proceed on each region independently, possibly with some interaction at the boundaries. It may be possible eventually to have a separate processor for each pixel, thus combining speed with flexibility. There are, however, some hidden problems with parallel processing. One of these is communicating data between processors. A system architecture could consist of a single master processor and a number of slaves. The master acts as a system controller passing packages of work (data) out to each slave. When the slaves have completed their tasks they signal to the master which collects the results (more data) and passes this on to, say, a machine control system. Unfortunately, most computers have a single pathway for communication between units, known as the bus. Increasing the number of processors on the bus will increase the speed of processing up to the point where the data traffic along the bus becomes too large. Thereafter, adding more processors will result in no extra performance and may even reduce it.

Parallel systems which can overcome the problem have been, until recently, very expensive. However, a significant development, although slow to gain popularity, has been the Transputer. This device was developed by the British company Inmos, now part of the French electronics group SGS Thomson. The central processor of a Transputer is a fairly conventional microprocessor but the unique feature of the device is that it contains four 'links' or serial channels along which it can communicate with other Transputers. In order to be effective these links must be fast. Transputer links normally operate at 20 Mbit/s, but devices with speeds of five times this rate have already been announced. Hence there is no computer bus; rather, each unit carries its own pathways for data communications so avoiding this particular bottleneck. By this method is should be possible to obtain a much more linear increase in speed with an increase in processor numbers. Fig. 6.2 shows a TRAM (Transputer Module) containing a T800 Transputer with 1 Mbyte of RAM external to the processor. The Transputer has within it 32 kbyte of fast RAM and incorporates a floating point unit for fast maths calculations. TRAMs are designed to plug into a motherboard holding typically eight or 16, and the motherboards can be held in a rack.

Fig. 6.2. A transputer module or TRAM.

A system recently installed at the Silsoe Research Institute is capable of holding 192 TRAMs in a box about 300 mm cube.

In order to solve a particular industrial problem it will be necessary to design an image analysis system to get a proper balance between cost, flexibility and effectiveness. However, it must be said that in the agricultural context, most image analysis problems at present are the subject of research rather than commercial development. An image analysis system for research has to fulfil different requirements. Probably the most important requirement of all is that it must be flexible. Almost by definition the user does not know exactly what the system will be called upon to do. It need not be capable of working at the full rate of an intended application as much image processing research is in algorithm development and most research scientists can assume that processing speeds will catch up with their requirements by the time the research can be commercialized. As will be seen below, image analysis in biological systems almost always requires specialized software to be developed. It is very important that a research system allows this to be done easily. A software library containing a range of commonly used image processing functions is extremely useful. Although these will not be optimized for a particular application they will enable a range of low level processing functions to be called upon without the chore of writing the code. An ideal situation is where the hardware manufacturer will supply a library complete with the source code for the subroutines. As well as being directly usable, these subroutines form the basis for new code to be developed by the user by modifying the library subroutines. A further useful facility, especially if the system is to be used to solve a particular problem in the short term, is that it should be compatible with a standard 'open' computer bus. Open, in this context, means that the computer system is meant to be designed by the user. This might involve choice of input/output devices, processor types, and memory size. It might mean specifying a multi-processor system with real-time operation to control a machine in which case the operating system must be capable of supporting this mode of operation. The Robotics and Control Group at Silsoe have standardized on the VME bus for their applications. Most work is done using the OS9 operating system and C as the programming language. Fig. 6.3 shows an image processing subsystem for this type of computer. It contains video interfaces for image capture and display, a framestore for holding images, and a microprocessor (Motorola

Fig. 6.3. Image processing subsystem.

68020) for image processing. The card is used within a VME bus environment using a master 68000 series processor for program development which has access to mass storage facilities (hard and floppy disks). Typically, programs are developed on the master processor under the OS9 operating system making use of the software library and by developing new algorithms. The developed program is transferred to the RAM on the image processing card and run either by a local OS9 system or as stand-alone code.

If necessary several such subsystems could be used on one VME bus. A particular example of this is a system recently installed at Silsoe for research into 3-D analysis of moving biological objects. Here the analysis is by matching stereoscopic pairs of images and deriving range information. It is important to acquire images at the same time, otherwise motion is confused with depth in the stereoscopic images. The system uses two cards of the type shown in Fig. 6.3 controlled by a master Motorola 68030 computer. Images are acquired simultaneously by each card and low-level processing such as edge extraction is done by the 68020 processors in parallel. A useful feature of these particular cards is that the framestore is accessible from the VME bus as well as by the local processors. Stereo matching is therefore done by the master 68030 processor accessing the two framestores via the bus.

Image analysis algorithms

Faced with the almost overwhelming high technology of the previous section, it is easy to gain the impression that everything in image analysis has been solved by electronics engineers. This is far from the truth. What we do have is a workable substitute for the human eye. In some respects the television camera may even be better, for example in seeing into the infra-red or at very low light levels, although the design is not nearly so elegant. Also we have electronic devices that can do arithmetic very quickly, many orders of magnitude faster than the human brain, and can store large amounts of data in a highly reliable way. Unfortunately, the electronics merely produces large quantities of numbers (grey level values), and although the computer can manipulate these numbers quickly there is no true intelligence in the process whatsoever.

Now let us turn to how an animal, say a human, does image analysis. Even a baby can recognize its parents. Presumably it does not store away numerical details of face shape and size along with the range of distortions and aspects a face can have. Even if it did, it would not be able to carry out any arithmetic on the measurements, let alone at the speed of a Transputer system. In fact face recognition (a good example of image analysis of a biological object) is something that we can do easily yet is an unsolved problem in image analysis. It is an interesting observation that if we wish to approach the effectiveness and flexibility in image analysis that a human possesses we are almost certainly going about it the wrong way. However, for the present, let us look at what an image analysis system can do.

Low-level processing

The distinction between low- and high-level processing is blurred, but generally low-level processing consists of fairly straightforward arithmetical manipulation of pixel data. The reader is referred to the work by Rosenfeld and Kak (1982) for an extensive treatment, but a few more commonly used techniques are briefly outlined here.

It is generally accepted that edges are very important features in a visual scene. These may be characterized by sharp changes in grey level or colour or texture. Detection of edges is therefore based on differencing. For example, an edge could be detected by scanning the image horizontally and when the difference between successive values exceeds a certain threshold an edge is recorded. Of course this would be sensitive to generally vertically disposed edges and may miss horizontal ones. The problem is overcome by differencing in both horizontal and vertical directions and combining the results. As with all signal processing, differencing leads to noise amplification and there is considerable scope for designing edge detectors while minimizing the effects of noise.

Edge detection is a filtering process. Another type of filter smooths the image; it is in a sense the opposite of an edge detector. Smoothing filters, which can reduce the effects of noise, are basically summing devices. A commonly used smoothing filter moves a 3×3 pixel window across the image. The centre pixel is set to the sum or the average of the 9 individual values within the window.

A further important technique is template matching which seeks to find a measure of similarity between an area of the image and a predefined prototype (the template). A template of grey levels is compared with parts of the image in turn. A measure of match is the sum of the products of grey level values of corresponding pixels in the image and the template.

Often it is important to identify areas of the image having similar properties, as these may come from a single physical object. Those separate parts of the image corresponding to objects are often called 'blobs'. If the objects have a characteristic grey level, say brighter than the background, and are not touching, a simple way of separating them is to threshold the image such that all grey values over the threshold are turned white and all values equal to or below the threshold are turned black. A second simple method is the chain code. In this technique, a grey level contour is followed. If we imagine the grey level distribution represented as a third dimension above the plane of the image, bright parts would be high and dark parts low – rather like a mountainous landscape. A chain code algorithm would be started on a contour and would follow it until it met with the starting point again. Chain coding is useful in a number of respects. If the object contrasts well with the background then the grey level contour can be chosen to represent the object edge. Also the form of the chain code is a set of numbers, each of which represents the direction from the current pixel on the contour to the next one. Thus a complete contour (or object boundary) is represented by a relatively short list of numbers, so achieving a large reduction in data compared with the original image. As might be expected there are various formulae available for calculating blob properties (such as area, perimeter, centroid, principal axis) directly from the chain code.

High-level processing

High-level processing is concerned with taking the features produced by low-level routines and using them to extract information about the scene. The standard work of reference is Ballard and Brown (1982). In the case of manufactured objects this part of processing can be trivial and image analysis is concerned mostly with operating the low-level routines fast enough. For example, consider a production line where jars of a food product are to be inspected to see whether a label is present. The jars contain a dark semi-liquid and the labels are a light colour. The transport mechanism carries a line of jars and stops each time a jar is in front of the camera. Because we know more or less exactly where a jar is we can concentrate on only that part of the field of view where we expect the label to be. A simple grey level threshold will determine the label part of the scene; we could chain code the blob representing the label and determine its centroid (is it in the correct place?), its shape (is is torn?), or its grey level texture (has it got writing on it?).

Now let us consider inspection of biological objects, say potatoes on a conveyor belt. This time we do not know in advance what the position of the objects will be when they are inspected, although we know the range of positions (they are on the belt somewhere). If the potatoes are fairly clean and the conveyor a dark colour we could possibly identify potato areas by thresholding. Unfortunately, unless we take the rather complicated mechanical step of separating potatoes, we don't know if a single blob is a single potato or a touching group and so it is difficult to draw conclusions about their size or shape. If we are dealing, say, with oranges we could assume that they were round and look for round objects in the scene. Unfortunately there is no such clear-cut mathematical model or template for a potato shape and so the problem is more difficult (although manageable – see the case study below).

It is this natural variability of biological objects that make their image analysis so challenging. A topic that has received considerable interest is the segmentation of man-made objects in natural scenes; examples are vehicles and military objects such as tanks and aeroplanes. One straightforward although time consuming technique is to employ a CAD (computer-aided design) type model of the object and present this as a template in various positions, orientations, and scales to the image. Matches are sought between image features, say edges or corners, and corresponding ones on the model. This can work well in constrained situations but is subject to many false or poor matches, especially if the edge map is incomplete or noisy as is typical in a natural scene, and also it takes a considerable time to search the 'space' of possible scales, positions and orientations. It may be possible to use some sort of model of a biological object but this would need to contain more degrees of freedom than its manufactured counterpart. For example, if the object were an animal, the model would have to cope not only with scale and orientation changes, but with distortions such as the relative motions of arms, legs and body. Obviously some advantage could be taken of constraints; for example, joints can move only in certain ways and at certain speeds. This approach has been used by Hogg (1983) in the segmentation of a walking man in a natural scene. Use of flexible templates has also been made by Widrow (1973), for the analysis of chromosome images, and by Tillett and Marchant (1990) for segmenting pigs in scenes.

Trainable classifiers

High-level image analysis can be regarded as a classification process. For example, given a set of features, one grouping is classified into component (a) − potato − and another into component (b) − conveyor belt. Or, at a higher level still, this pattern of grey levels represents a pig's head and this other pattern a pig's rump. It is useful here, for the purpose of illustration, to return to the face-recognition problem above. Most of us can walk into a crowded room and associate faces with names (that is, recognize people). If we were challenged to write an image analysis algorithm to do this, which involves reducing the recognition process to numerically based rules with the grey levels of the image as data, we would find it impossible. In a more agricultural context we might wish to classify the gait of an animal to determine lameness or the pattern of surface markings on fruit to determine soundness. In all cases it would be difficult to say 'how' we classify (in numerical terms), but relatively easy to say 'what' we classify, that is to make the decision. In other words, many image analysis tasks are subjective and this means it is difficult to write a computer program to automate them.

One way around this difficulty is to use a trainable pattern classifier. With this technique an explicit classification algorithm is not programmed into the computer; rather, the machine is presented with a large number of examples of data and 'learns' to classify it into various categories. This can be done either in an unsupervised mode − the computer will recognize certain groupings of features in the data itself and classify subsequent presentations of data into these groupings – or with a supervisor – a human interacts with the machine when the training data is presented and classifies it; when presented with data subsequently the machine will classify automatically. In the most general case it is not necessary even to tell the computer what the important features of classification are. However, this knowledge will generally result in more effective classification or shorter training times. For example, if it were necessary to classify the ends of a generally sausage-shaped blob into either an animal's head or rump, it would be useful to tell the program that an important feature was the smoothness of the outline (the head outline is generally less smooth than the rump). A simple type of trainable classifier, the WISARD (Stonham, 1986), has been used by Onyango *et al.* (1989) to recognize junctions between plant stems and sideshoots for automatic propagation. This classifier has also been quite successful in the face-recognition problem mentioned earlier.

What is possibly the most exciting prospect at present is a biologically inspired computing method called a neural network (Wasserman, 1989). Very little is understood about the construction and operation of animal brains but it is possible to observe large numbers of specialized cells called neurons in their make up. These neurons typically have several inputs, which receive signals from other neurons, and one output path. Activity can be observed on the output path when sufficient stimuli appear on the inputs. Whether a neuron has an active output depends on the total stimulation of its inputs but each input can have a different importance or 'weight'. Also the output appears to switch on or off depending on whether the total stimulation exceeds some threshold. It is perhaps too early to observe that human knowledge and intelligence may be stored in the way these neurons are connected together and in the weights of the connections, but such observations have inspired researchers to build electronic analogues or computer simulations of neurons and connect them as a network. The surprising feature of these neural networks is that they show an ability to learn. This is generally done by presenting the network with a set of training patterns – in the agricultural context this could be, for example, a set of potatoes having different shapes (Tillett *et al.*, 1989), along with a human classification of these patterns. The network gradually adjusts the weights between interconnections, following a predefined method, thus storing knowledge about the patterns and the classes into which they fall. From then on the network can classify patterns it has never seen before. This ability to generalize seems to fit in with a tentative definition of intelligence, lacking in conventional computer programs – that is, the ability of a machine to respond in a sensible way to a situation it has never encountered before.

Neural networks have received attention for some considerable time – since the late 1940s. Even with the advent of computers, interest declined after it was proved that a certain type of network could not handle a class of problems containing some very simple examples. However, in more recent years more powerful networks have been developed along with methods of training them and the re-awakening of interest has been almost explosive. Currently most research is done using simulations of networks on conventional serial processors. This training process can be very time consuming, but recent developments in parallel processors, special devices such as vector processors, and even dedicated neural chips should see training time reduce rapidly.

Applications in agriculture

This section is not meant to be an exhaustive literature survey but is intended to indicate the possible range of application of image analysis to agriculture and food production. More complete reviews can be found in Marchant (1990a) and Tillett (in press).

To gain an idea of possible applications we only need to ask the question 'When do we use our eyes?' Most applications fall into two categories, guidance and inspection. In guidance applications the image processing is used to direct the movement of a robotic manipulator or a vehicle. Current or recent research includes the harvesting of oranges (Harrell *et al.*, 1989; Slaughter and Harrell, 1989), tomatoes (Kawamura *et al.*, 1987; Whittaker *et al.*, 1987), mushrooms (Tillett and Reed, 1990), apples (Sites and Delwiche, 1988; Grand d'Esnon *et al.*, 1987), and cucumbers (Shono *et al.*, 1989); also automating plant propagation (Tillett *et al.*, 1990), the transfer of bacterial colonies (Jasiobedzki and Martin, 1989), robotic butchery (Khodabandehloo and Brett, 1990), and guiding farm machinery (Reid and Searcy, 1988). Inspection or grading applications include oranges (Maeda, 1987), potatoes (McClure and Morrow, 1987; Marchant *et al.*, 1990), apples (Davenel *et al.*, 1988; Rehkugler and Throop, 1989), carrots (Batchelor and Searcy, 1989), peppers (Wolfe and Swaminathan, 1987), tomatoes (Sarkar and Wolfe, 1985), biscuits (Davies, 1984), meat (Cross *et al.*, 1983; Keller *et al.*, 1986), oysters (Awa *et al.*, 1988), pot plants (Hines *et al.*, 1987), strawberry plants (Cardenas-Weber *et al.*, 1988), and tree seedlings (Rigney and Kranzler, 1988). There has also been some work on seed inspection for corn (Gunasekaran *et al.*, 1987), rice (Sistler, 1990), and soybeans (Paulsen *et al.*, 1989). A very challenging area is the application of image analysis to live animal situations such as the sizing of pigs (Schofield and Marchant, 1990) and fish (Caputo *et al.*, 1990).

Some of these applications are likely to become practicable much sooner than others. The more constrained the situation the more likely is early realization. For example, grading vegetables where the presentation and lighting is controlled has already been solved to some extent (see the case study below). However, there are still significant problems to overcome even here. For example, if we wish to match human inspectors dealing with, say, apples, it will be necessary to tell the difference between dark skin areas which could be blemishes, and others which could be stalk or calyx. Algorithm reliability is also a problem. If a packhouse pays a large sum for a high-technology grading machine it must grade consistently virtually all of the time. It is relatively easy to obtain, say, 80% success in automatic inspection, but rates above 99% (necessary to achieve commercial acceptance) require a great deal of artificial intelligence to be developed and included in the software.

In less constrained situations, such as harvesting or animal behaviour classification, much more time and research will be needed before systems become practically feasible. This time will be necessary to develop the artificial intelligence required to match even a fraction of human perception; it may also require considerable research into how we, as humans, achieve the recognition and classification ability that we have.

Case study – potato grading

As the expectations of the buying public rise there is an increasing need for shops and supermarkets to offer top quality produce. This means conformity to an ideal shape and size and freedom from defects. Grading of produce to ensure this conformity is very largely done by human inspectors where the produce is presented on some sort of moving belt. As with all human inspection tasks, performance varies between inspectors and over time causing loss of value in the product as good specimens are rejected, and wastage of shelf space and customer dissatisfaction as bad specimens are classed as good.

One requirement for this work was to select potatoes for baking in their jackets. In order to be of baking quality, specimens must be of a certain minimum size, both overall and in their three principal axes. Also they must not be an 'ugly' shape, the ideal generally being taken as a prolate spheroid. A further requirement (the solution, however, not reported here) was to inspect the surface of the potatoes and record sizes and numbers of atypical regions. These regions could then be regarded as blemishes if over a certain size or number or of a certain colour, and the machine instructed to grade accordingly.

Despite the fact that there is a considerable premium on the value of baking-quality potatoes, in absolute terms they are still low-value items. This means that, in order to make economic sense, inspection speeds must be high, a figure of about 40 potatoes per second being acceptable. Other constraints on the system design arose because the project was funded by private industry. Firstly, the work had to be completed within one year. This meant that off-the-shelf existing technology had to be used. Secondly, the precise requirements for sorting could be changed both during the development phase and afterwards. This implied a high degree of flexibility in the system, so most of the processing was done in software using hardware subsystems only where absolutely necessary.

Mechanical arrangement

The area in which the potatoes are presented to the television camera is shown in Fig. 6.4(a), and the grading-out section in Fig. 6.4(b). More details are given in Marchant (1990b), but the important points are:

1. the potatoes rotate during their forward travel, so exposing most of the surface to the camera; and
2. although the potatoes are formed into rows by the roller mechanism, no attempt is made to singulate the potatoes across the grading belt.

The attributes that need to be measured are length, minimum width, maximum width (if the potatoes were spheroids these would be the three principal axes), shape, and an estimate of the weight. This last attribute is obtained from a volume estimate which, in turn, is obtained from a number of projected area measurements. In order to get a good estimate of the widths and the volume, a number of views of each potato have to be taken. This number has been fixed at 12. This, along with the need to inspect the surface of the potatoes, is the reason for rotating them.

Avoiding singulation has a number of advantages. Firstly, the conveyor and its feed system can be simpler. It is extremely difficult to feed objects (especially if these are natural ones having no uniform shape or size) into cups on a conveyor at high speed. Typical problems are objects bouncing out of cups, or having more than one object in a cup. Secondly, there can only be a fixed number of cups across the conveyor. The number is determined by the largest object expected. With an average distribution of object sizes the conveyor must run at significantly reduced throughput. However, singulation can only be avoided if the system can identify the position of individual objects and store this information until the time is right to operate the grading mechanism. Identification must include the delineation of individual objects when they form touching groups. This task is made more difficult with natural objects like fruits or vegetables as opposed to, say, golf balls where the size and shape is known beforehand.

Lighting within the viewing cabinet is even and controlled but even here there are problems. Although the rollers are black, their smooth plastic surface can give rise to reflections and the resulting bright areas could be taken to be potatoes. This is overcome by using polarizers on the lighting banks with a crossed polarizer on the camera lens. In specular reflection (from

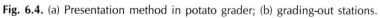

Fig. 6.4. (a) Presentation method in potato grader; (b) grading-out stations.

the smooth rollers) the polarization is preserved and so the light is blocked by the lens polarizer. The diffuse reflecting surface of the potatoes destroys the polarization, and light from the potatoes therefore passes through to the camera lens.

Image analysis

Leaving out the problems involved with blemish detection, there were two major image analysis difficulties: the first was dealing with the data quickly enough, and the second was delineating individual potatoes in touching groups reliably.

Speed of processing has been dealt with in some detail elsewhere (Marchant *et al.*, 1988) and the following is a brief description only. It was assumed that approximately ten potatoes could be accommodated on each roller, which gave a conveyor speed of four rollers per second at full capacity. When the optical system was adjusted to take in the whole width of the belt, approximately four lines of potatoes were in view so it was decided to use three of these for processing in any one view. At 12 views per potato, this gave a processing rate of 16 video frames per second. It was decided that the minimum resolution to give adequate performance was 256 pixels square, thus a pixel rate of just over one million per second was required.

As this part of the system required knowledge only of the potato outline, the first stage of processing was thresholding followed by boundary encoding. Thresholding was done by using a simple comparator directly on the analogue video signal and taking no processor time. Boundary encoding was done using a hardware subsystem specially designed for the task − this was a considerable design exercise and the result occupies a 200 mm square seven-layer printed circuit board. A hardware system was necessary in order to achieve the required speed: the unit can encode the boundary of an average sized potato in about 200 μs and the single unit operates on all three rows of potatoes under analysis. As mentioned previously, boundary encoding can reduce the amount of data considerably, in this case from 64 to about 3 kbytes. The decision to use a special hardware-based data reduction unit was taken with great care. Although hardware units are fast, they are also inflexible. However, it was felt that boundary encoding would always be required whatever produce was used, so the potential inflexibility would not be a problem, all produce-specific processing being done in software.

After data reduction, a parallel processing technique was used to analyse the remaining data. Three slave processors (Motorola 68010) were used, each containing the same program and each operating on a single row of potatoes. After a row passed out of the field of view that particular processor was freed and able to pick up the next row coming into the viewing area. A master processor (Motorola 68000) was used to pass chain code data from the hardware unit to the slaves, and also to collect measured attributes (lengths, widths, etc.) from the slaves. These data were passed to a separate processor to operate the mechanical parts of the machine. The master processor also handled human interaction with the machine as a separate task, so enabling the alteration of settings while the machine was running. Hence much use was made of the multitasking facilities of the OS9 operating system, along with its ability to handle real-time events – for example, synchronization of the programs with the mechanical operation via interrupts. Colour image analysis and blemish detection was handled by a fourth slave processor. A useful side-effect of the multi-processor architecture was that it enabled different teams to work largely independently. The only part requiring close coordination being the protocols for data transfer defined at the interfaces between each subsystem.

The second major problem was delineating individual potatoes and ascribing attributes to the correct ones. Once again, this has been covered in detail elsewhere (Marchant *et al.*, 1990) and only a brief summary is given here. A characteristic of the point of contact between two touching potatoes is a pair of cusps in the outline approximately opposite each other. These cusps can be found using standard methods of detecting corners. As with all image processing for biological objects, it is not possible to define the parameters of the cusp exactly and a balance has to be struck between detecting most true cusps along with some false ones, and detecting only true cusps at the expense of missing a large number. It was found to be impossible to detect all true cusps even when a large number of false ones was allowed. However, noting that true touching points generally consist of two opposing cusps it was possible to obtain a reasonable balance. This method of delineation was found to be effective about 99% of the time. At first this seems adequate until three points are considered. Firstly, at the full speed of operation there would be a misgrading every 2.5 s. Secondly, wrong delineation may well affect a whole row of potatoes rather than a single one. Thirdly, the tests (like most machinery tests) were carried out away from a processing factory – all practical engineers should expect a down-grading of performance under real conditions, for example when the conveyor belt or optics are very dirty and when the designer of the machine is not present!

Thus, improved performance was required. Here advantage was taken of the many views gained of each row of potatoes. If we assume for the moment that each view is uncorrelated with respect to the difficulty of delineation and a failure occurs in 1% of cases, then if we take, say, six views and combine the results, the likelihood of failure in all six is $(0.01)^6$, in other words very small indeed. In fact the views are correlated as they contain the same potatoes in the same order across the belt, but tests using a combination technique failed to show any failures when 1,000 potatoes were passed over the belt.

Underlying this example, there is a general principle which is very useful in image processing of biological objects. It is generally not possible to get a single algorithm to work perfectly with a biological object because of the inherent variability. If several algorithms based on different principles are used, or the same algorithm under different conditions, the resulting combination should increase reliability. This general principle takes advantage of the redundancy in the set of data; that is, it should be possible to obtain a correct result with only one algorithm – the others are redundant but are used as a safety net.

The foregoing example shows that it is possible to solve a practical problem using image analysis. It can, however, take considerable effort of a highly technological nature, the techniques used being based on knowledge gained during a programme of basic underpinning research.

Conclusions and future prospects

Image analysis is an important technology that will find many applications in modern engineering. Image analysis equipment is already extremely powerful and will continue to improve, following development in computing technology. However, while the sensory and central processing elements are well advanced, the ability to emulate human intelligence in handling visual data is at a very early stage indeed. Image analysis in well-structured environments containing manufactured objects is reasonably straightforward. However, in the contexts of agriculture and food production, the objects of interest are always variable and are often found in environments which cannot be controlled. This sets image analysis in biological systems apart: while some problems can be solved now, others will require a great deal of research. A major research area is how to encapsulate, represent, and implement an artificial version of human intelligence. If this can be done many processes which have defied automation will become the subject of automatic systems using image analysis as the major method of control.

References

Awa, T. W., Byler, R. K. and Diehl, K. C. (1988) Development of an inexpensive oyster meats grader. Paper no. 88-3539. American Society of Agricultural Engineers, St Joseph MI.

Ballard, D. H. and Brown, C. M. (1982) *Computer Vision*. Prentice Hall, Englewood Cliffs NJ.

Batchelor, M. M. and Searcy, S. W. (1989) Computer vision determination of the stem/root joint on processing carrots. *Journal of Agricultural Engineering*

Research 43, 259–69.

Caputo, J., Johnson, A. and Mayr, N. (1990) Let me count the ways. *Canadian Aquaculture* 6, 47–50, 56.

Cardenas-Weber, M. C., Lee, F., Guyer, D. E. and Miles, G. F. (1988) Plant features measurements with machine vision and image processing. Paper no. 88-1541. American Society of Agricultural Engineers, St Joseph MI.

Cross, H. R., Gilliland, D. A., Durland, P. R. and Seideman, S. (1983) Beef carcass evaluation by use of a video image analysis system. *Journal of Animal Science* 57, 908–17.

Davenel, A., Guizard, C., Labarre, T. and Sevila, F. (1988) Automatic detection of surface defects on fruit by using a vision system. *Journal of Agricultural Engineering Research* 41, 1–9.

Davies, E. R. (1984) Design of cost effective systems for the inspection of certain food products during manufacture. *Conference Proceedings – 4th International Conference on Robot Vision and Sensory Controls*. British Robot Association, London, 437–46.

Grand d'Esnon, A., Rebatel, G., Pellenc, R., Journeau, A. and Aldon, M. J. (1987) MAGALI: a self propelled robot to pick apples. Paper no. 87-1037, American Society of Agricultural Engineers, St Joseph MI.

Gunasekaran, S., Cooper, T. M., Berlage, A. G. and Krishnan, P. (1987) Image processing for stress cracks in corn kernels. *Transactions of the American Society of Agricultural Engineers* 30, 266–71.

Harrell, R. C., Slaughter, D. C. and Adsit, P. D. (1989) A fruit tracking system for robotic harvesting. *Machine Vision and Applications* 2, 69–80.

Hines, R. L., Sistler, F. E., Wright, M. E. and Brown, L. (1987) Establishing grading standards for container grown plants. Paper no. 87-6053. American Society of Agricultural Engineers, St Joseph MI.

Hogg, D. (1983) Model-based vision: A program to see a walking person. *Image and Vision Computing* 1, 5–20.

Jasiobedzki, P. and Martin, W. J. (1989) Processing of bacterial colony images for automatic isolation and transfer. *Journal of Physics E: Scientific Instrumentation* 22, 264–7.

Kawamura, N., Namikawa, K., Fujiura, T. and Ura, M. (1987) Study of fruit harvesting robot and its application to other works. *Conference Proceedings – Agricultural Machinery and International Cooperation in High Technology Era*. University of Tokyo, Tokyo, 132–8.

Keller, J. M., Covavisaruch, N., Unklesbay, K. and Unklesbay, N. (1986) Color image analysis of food. *Conference Proceedings – Computer Vision and Pattern Recognition*. IEEE Computer Society, Miami Beach, FL, 619–21.

Khodabandehloo, K. and Brett, P. N. (1990) Intelligent robot systems for automation in the food industry. *Conference Proceedings – Mechatronics: Designing Intelligent Machines*. Institution of Mechanical Engineers, Cambridge, UK, 247–54.

McClure, J. E. and Morrow, C. T. (1987) Computer vision sorting of potatoes. Paper no. 87-6501, American Society of Agriculture Engineers, St Joseph MI.

Maeda, H. (1987) Grade and color sorter through image processing. *Conference Proceedings – Agricultural Machinery and International Cooperation in High*

Technology Era. University of Tokyo, Tokyo, 350–6.

Marchant, J. A. (1990a) Computer vision for produce inspection. *Postharvest News and Information* 1, 19–22.

Marchant, J. A. (1990b) A mechatronic approach to produce grading. *Conference Proceedings – Mechatronics: Designing Intelligent Machinery.* Institution of Mechanical Engineers, Cambridge, UK, 159–64.

Marchant, J. A., Onyango, C. M. and Street, M. J. (1988) System architecture for high speed sorting of potatoes. *Conference Proceedings – Automated Inspection and High Speed Vision Architectures II.* Society of Photo-optical Instrumentation Engineers, Cambridge MA, 171–3.

Marchant, J. A., Onyango, C. M. and Street, M. J. (1990) Computer vision for potato inspection without singulation. *Computers and Electronics in Agriculture* 4, 235–44.

Onyango, C. M., Marchant, J. A. and Street, M. J. (1989) A simulation of the WISARD adaptive image classifier in 68000 FORTH. *Computers and Electronics in Agriculture* 4, 129–38.

Paulsen, M. R., Wigger, W. D., Litchfield, J. B. and Sinclair, J. B. (1989) Computer image analysis for detection of maize and soybean kernel quality factors. *Journal of Agricultural Engineering Research* 43, 93–101.

Rehkugler, G. E. and Throop, J. A. (1989) Image processing algorithm for apple defect detection. *Transactions of the American Society of Agricultural Engineers* 32, 267–72.

Reid, J. F. and Searcy, S. W. (1988) An algorithm for separating guidance information from row crop images. *Transactions of the American Society of Agricultural Engineers* 31, 1624–32.

Rigney, M. P. and Kranzler, G. A. (1988) Machine vision for grading southern pine seedlings. *Transactions of the American Society of Agricultural Engineers* 31, 642–6.

Rosenfeld, A. and Kak, A. C. (1982) *Digital Picture Processing*, 2nd edn. Academic Press, Orlando FL.

Sarker, N. and Wolfe, R. R. (1985) Feature extraction techniques for sorting tomatoes by computer vision. *Transactions of the American Society of Agricultural Engineers* 28, 970–4.

Schofield, C. P. and Marchant, J. A. (1990) Image analysis for estimating the weight of live animals. *Conference Proceedings – Optics in Agriculture.* Society of Photo-Optical Instrumentation Engineers, Boston MA.

Shono, H., Amaka, K. and Takura, T. (1989) Detection of cucumber fruit position by image processing. *Journal of Agricultural Meteorology* 45, 87–92.

Sistler, F. E. (1990) Grading agricultural products with machine vision. *Conference Proceedings – IEEE International Workshop on Intelligent Robots and Systems.* IEEE, Tsuchiura, Japan, 255–61.

Sites, P. W. and Delwiche, M. J. (1988) Computer vision to locate fruit on a tree. *Transactions of the American Society of Agricultural Engineers* 31, 257–63, 272.

Slaughter, D. C. and Harrell, R. C. (1989) Discriminating fruit for robotic harvesting using color in natural outdoor scenes. *Transactions of the American Society of Agricultural Engineers* 32, 757–63.

Stonham, T. J. (1986) Practical face recognition and verification with WISARD. In:

Ellis, Jeeves, Newcombe and Young (eds), *Aspects of Face Processing*, Martinus Nijhoff, 426–41.

Tillett, R. D. (in press) Image analysis for agricultural processes: a review of potential opportunities. *Journal of Agricultural Engineering Research*.

Tillett, R. D. and Marchant, J. A. (1990) Model-based image processing for characterizing pigs in scenes. *Conference Proceedings – Optics in Agriculture.* Society of Photo-optical Instrumentation Engineers, Boston MA.

Tillett, R. D. and Reed, J. N. (1990) Initial development of a mechatronic mushroom harvester. *Conference Proceedings – Mechatronics: Designing Intelligent Machines.* Institution of Mechanical Engineers, Cambridge, UK, 109–14.

Tillett, R. D., Brown, F. R., McFarlane, N. J. B., Onyango, C. M., Davis, P. F. and Marchant, J. A. (1990) Image guided robotics for the automation of micropropagation. *Conference Proceedings – International Workshop on Intelligent Robots and Systems.* IEEE. Tsuchiura, Japan, 265–70.

Tillett, R. D., Onyango, C. M., Davis, P. F. and Marchant, J. A. (1989) Image analysis for biological objects. *Conference Proceedings – 3rd International Conference on Image Processing and its Applications.* Institution of Electrical Engineers, Warwick, 207–11.

Wasserman, P. D. (1989) *Neural Computing Theory and Practice.* Van Nostrand, New York.

Widrow, B. (1973) The 'rubber mask' technique – I: pattern measurement and analysis. *Pattern recognition* 5, 175–97.

Whittaker, A. D., Miles, G. E., Mitchell, O. R. and Gaultney, L. D. (1987) Fruit location in a partially occluded image. *Transaction of the American Society of Agricultural Engineers* 30, 591–6.

Wolfe, R. R. and Swaminathan, M. (1987) Determining orientation and shape of bell peppers by machine vision. *Transactions of the American Society of Agricultural Engineers* 30, 1853–6.

Chapter 7

Climate Modelling and Control in Greenhouses

Bernard Bailey

The greenhouse climate

The function of a greenhouse is to provide an environment in which plants can be grown in a cost-effective way. It gives protection from the adverse effects of wind and the various forms of precipitation, and, because air exchange is restricted, above-ambient temperatures are produced during the day by the solar radiation absorbed inside the greenhouse. However, at night and during winter the temperature which occurs naturally in greenhouses is only slightly above ambient and the plants derive less benefit from being enclosed. It has been known for almost as long as greenhouses have been used that it is beneficial to provide artificial heating to prevent the greenhouse temperature from becoming too low. During summer when the converse occurs and the temperature becomes too high, it is necessary to cool the greenhouse by providing ventilation.

The factor which has the largest influence on plant growth and development is solar radiation, as this provides the energy for photosynthesis, the basic mechanism of plant growth. The irradiance inside a greenhouse is lower than outside because of losses at the cover. The natural variability of solar irradiance causes variations in the rate of plant growth, and in winter growth is limited by low irradiance. Artificial illumination is used to increase growth rates and improve the quality of some plants, notably during plant propagation and in the production of flowers, but it is too expensive for widespread use in production greenhouses. To maximize productivity greenhouses are designed to have the highest transmission of photosynthetically active radiation which is compatible with providing a structure of adequate strength. During photosynthesis, carbon, obtained from carbon dioxide in the greenhouse air, is assimilated by the growing plants. As the greenhouse is essentially a closed system this lowers the concentration of carbon dioxide, which in turn reduces the rate of photosynthesis and consequently the rate of plant growth; to avoid this, carbon dioxide is frequently added to the greenhouse.

Only 2–3% of the solar radiation absorbed by plants is used in photosynthesis; the remainder is transferred to surrounding surfaces by thermal radiation, to the air by convection or is used as latent heat to evaporate water in transpiration from leaves. The latter process is of considerable significance in limiting leaf temperature when insolation is high, and the resulting flow of water through the plant plays an important role in the uptake and distribution of nutrients. However, the water vapour produced raises the moisture content of the greenhouse atmosphere above the value in the external air, and in well-sealed energy-efficient greenhouses, the humidity can become so high that transpiration is reduced, adversely affecting plant growth. If transpiration ceases then the possibility exists that moisture will condense on leaves providing conditions conducive to the development of fungal diseases. Consequently it is necessary to prevent the humidity of the greenhouse air from becoming too high. However, low humidity is also harmful because high transpiration rates can be detrimental, especially to young plants.

The requirements of the cover to allow the transmission of short-wave solar radiation and to isolate the greenhouse interior from the weather conflict. The compromise solution means the greenhouse is a relatively responsive system directly affected by the outside conditions. In general these conditions are highly dynamic, both on short and long timescales. The short-term effects result mainly from the variations in insolation experienced under partially cloudy skies and occur over periods of minutes, while the long-term effects are caused by daily and seasonal fluctuations. Climate control is used in the commercial production of crops in greenhouses to reduce the influence of the weather, to provide conditions in which plant growth and development is more predictable and to ensure the fruit, vegetables and flowers produced are of high quality.

The development of climate control

The major climatic variables over which it is practicable, and advantageous, to exercise control are temperature, and the concentrations of carbon dioxide and water vapour in the greenhouse air. Temperature is the most important of these and has received the most attention. Early heating systems consisted of stoves with flues, either passing through the greenhouse or built into its walls, distributing the heat. This rather uncontrolled form of heating was replaced by hot water heated in furnaces and then passed through large-bore cast-iron pipes in the greenhouses under the influence of the thermosyphon created by differences between the densities of cold and hot water. Control was provided manually by varying the rate of stoking of the coal-fired furnace. The development of pumped circulation systems using 35–50 mm bore steel pipes containing water heated in oil-fired boilers provided the opportunity for much closer control over greenhouse air temperature because of the reduced thermal capacity and the faster response. Electronic analogue controllers, providing feedback control with proportional and integral control actions, were introduced, using aspirated thermometers shielded from radiation, to control the supply of heat to individual greenhouses. These controllers permitted minimum temperatures to be maintained in greenhouses which enabled the most suitable day and night 'blueprint' temperatures for important greenhouse crops to be determined experimentally. Some controllers provided light-dependent control of temperature, and raised the heating temperature set-point during the day with increasing irradiance; others enabled the night temperature to be related to the radiation integral of the preceding day.

Electronic controllers were also developed to operate ventilators in the greenhouse roof to provide cooling during summer. These used a variety of forms of control including pulse floating, proportional control and a combination of both. Wind direction was sensed to ensure the leeward ventilators would open first, with those on the windward side only being used if necessary. Wind speed was also monitored in order that the ventilators could be closed during high winds to prevent damage to the greenhouse.

The heating and ventilator controllers were linked in order to provide a way of limiting the humidity of the greenhouse air. The ventilators were opened to allow the humid warm air to escape to be replaced by drier, but cooler, outside air, and heating was used to maintain the desired air temperature. While ventilators were being used to control humidity, limits were set on the permitted ventilator opening to avoid wasting energy by excessive heating.

As a result of these increasingly sophisticated forms of control being required by growers, control systems were becoming very complex. The technical difficulties of controlling a large number of variables, some of which had to be coupled, were reduced with the introduction of digital controllers based on microcomputers. These had the capability of providing integrated control over a large number of variables and implementing complex control algorithms at a cost which compared favourably with analogue controllers. The digital controllers gave immediate improvements in monitoring the conditions which occurred in the greenhouses, the ease of changing the set-points of the controlled variables, the reproducibility of the control signals over long periods, and the availability of alarm functions. Also, by linking a meteorological station to the controller some decisions on set-point changes could be made automatically.

It was soon realized that the computing power available offered possibilities for making substantial improvements in greenhouse climate control. Classical feedback control, based on detecting changes in the greenhouse environment, could be augmented by feedforward control, in which models of the system are used so that the effects of changes in conditions outside the greenhouse can be predicted before the internal conditions are affected, thus improving the quality of control. Controlling one variable, such as temperature, can influence other climatic factors, such as the concentrations of carbon dioxide and water vapour in the greenhouse atmosphere. The overall performance of climate control could be improved if the interactions between these coupled variables were explicitly included in the control algorithms. This is made possible by using multivariable control, which is feasible with computer-based control systems. Also, the door has been opened to the prospect of increasing the effectivenesss of climate control by the introduction of optimization techniques. The optimum climate is that which gives the largest margin between the value of the crop produced and the cost of creating the climate.

Advances in these areas require information on how the greenhouse climate is influenced by changes in external conditions and by the presence of a crop, and on how plants respond to the greenhouse climate. This information is becoming available in the form of models which describe various parts of the greenhouse system. Physicists and engineers have been developing both static and dynamic physical models to describe the greenhouse climate and plant scientists have been creating models aimed at describing the growth, development and production of plants.

Greenhouse climate models

The essential requirement of models of the greenhouse climate is to define how the conditions inside the greenhouse are influenced by the outside weather, the nature of the greenhouse, and the climate control system. Models must therefore describe the penetration of solar radiation into the greenhouse, the heat and mass balances of the elements which constitute the greenhouse system, and the characteristics of the climate control systems. The physical behaviour of the crop must be included, as it plays a crucial role in absorbing solar radiation and establishing the thermal and water balances of the greenhouse. Attempts have been made (Nara, 1979; den Hartog, 1988) to represent the spatial distribution of the climate on a two- or three-dimensional grid and to solve the set of energy, mass and momentum balances at the grid points, but only very simplified greenhouse geometries can be dealt with because of the limited capacity of computers. Therefore the problem is usually simplified by considering only the static behaviour of the greenhouse, or by considering a large greenhouse in which edge effects can be neglected and the problem reduced to one with only a single dimension. When the time-dependent behaviour of the greenhouse is of interest a detailed representation of the spatial distribution of climate is not necessary as attention is usually focused on the response of the spatial average climate.

Steady state models

Early studies of the greenhouse climate concentrated on determining the thermal behaviour of the greenhouse. The basic responses to the external weather were determined experimentally by Whittle and Lawrence (1959, 1960). Businger (1963) identified and formulated many of the energy fluxes in the greenhouse and proposed one of the first energy balance models of a greenhouse. This type of model was developed further by Walker (1965) and used to determine the temperatures in a ventilated greenhouse. Morris and Winspear (1967) incorporated a mass balance for water and used the resulting model to study the influence of heating and ventilating on greenhouse humidity. Their approach was limited by the difficulty of solving the complex set of simultaneous equations.

More comprehensive models were subsequently developed based on energy and water-mass balances for the crop–greenhouse system which consisted of the crop, air, soil and greenhouse cover (Fig. 7.1). The balance equations were formulated in terms of the energy and mass fluxes relevant for each element, which in turn were defined in terms of the state variables of temperature and water vapour pressure. The modes of heat transfer included were thermal radiation, natural and forced convection, conduction, and latent heat. The boundary conditions usually consisted of the outside air temperature, absolute humidity, solar radiation, long-wave sky radiation, wind speed, and the soil temperature at a specified depth beneath the greenhouse. Other inputs to the models were the physical characteristics of the crop and the greenhouse. The balance equations, with an appropriate set of boundary conditions, were solved numerically using computers. Models of this type have been developed by Garzoli and Blackwell (1981), Baille *et al.* (1985), Jolliet *et al.* (1985) and Short and Breuer (1985), and have been used primarily to determine the heat loss from greenhouses. By using them with averages of meteorological data the energy required to grow specific crops could be obtained.

Steady state models have been used widely to study the behaviour of particular elements in the greenhouse and to assess the potential for new technologies. Tantau (1975) used a static model to investigate the reduction in energy consumption resulting from covering greenhouses with high thermal resistance materials. The effectiveness of using thermal screens to insulate greenhouses during the night, and the influence of the screen properties on

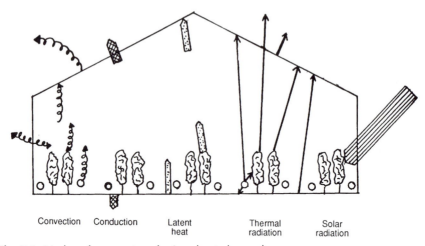

Convection   Conduction   Latent   Thermal   Solar
                              heat   radiation   radiation

Fig. 7.1. Modes of energy transfer in a heated greenhouse.

their performance, were similarly studied by Bailey (1981) and Okada (1985). Static energy and mass balance models were used by Bailey (1984) and by Seginer and Kantz (1989) to investigate the energy requirements of various ways of dehumidifying greenhouses.

The simpler formulations of greenhouse models contain a number of parameters, such as the overall heat transfer coefficient, the proportion of solar radiation converted to sensible heat, and the effective thermal capacitance of the greenhouse elements. The technique of *in situ* calibration has been used by Albright *et al.* (1985), Seginer and Kantz (1986) and Seginer *et al.* (1988) to obtain values for these parameters. Garzoli and Blackwell (1987) used a similar technique to determine the convective heat transfer coefficient in the space between two layers of plastic film covering a tunnel greenhouse.

Time-dependent models

The value of static models begins to diminish when the response time of the greenhouse becomes comparable with the rates of change of the boundary conditions. The time constants relevant to the important elements described by the model can be approximated by the ratio of the thermal capacity of the element to the overall heat transfer coefficient. Typical values for the crop, air, greenhouse cover, and the top 1 cm of the soil lie in the range 2 to 5 min; for thick layers of soil the thermal capacity increases while the transfer coefficient decreases; and the time constant increases to hours at a depth of about 5 cm, and to days at 15 cm. The response time of heating systems varies from minutes for air heaters, to hours for cooling hot water pipes. Thus when the soil and piped heating systems are considered a time-dependent model is necessary.

The techniques developed for analysing the time-dependent response of greenhouses can be divided into two groups: analytical methods and numerical methods. The analytical techniques require the assumption of linearity, and solutions were derived using weighting function techniques (Takakura, 1967, 1968), by the passive network technique (Froehlich *et al.*, 1979), and by using analogue computers (O'Flaherty *et al.*, 1973). However, the rapid development of digital computers and the introduction of simulation languages such as CSMP, resulted in the numerical approach becoming dominant. Takakura *et al.*, (1971) developed a greenhouse model which was solved numerically. In this model the flow of heat in the soil was treated as being two-dimensional. A number of similar models were subsequently developed. Although the basic structures were similar, each addressed particular aspects of greenhouse behaviour and different numerical integration methods were used. Models of particular note were those of Kimball (1973, 1981), Soribe and Curry (1973), Chandra *et al.* (1981), van Bavel *et al.* (1981), Avissar and Mahrer (1982), von Elsner (1982), Bot (1983), Cooper and Fuller (1983) and Deltour *et al.* (1985). These models enable the physical conditions in the greenhouse to be represented reasonably accurately, as shown by Fig. 7.2.

The models of Takakura *et al.* (1971), van Bavel *et al.* (1981) and Bot (1983) were compared by using each to calculate the energy required for heating and cooling using the same data sets (van Bavel *et al.*, 1985). The major differences between the models related to the way of simulating the penetration and absorption of solar radiation, the treatment of heat and mass transfer between the crop and the greenhouse air, the description of stomatal behaviour, the control of heating, ventilation and cooling, and in the way of estimating long-wave sky radiation. On the whole, the predictions of air temperature, humidity, and heating requirements were in good agreement, but there were significant differences between the predictions of transpiration and the requirement for evaporative cooling. The main conclusion was that knowledge was lacking about the convective exchange processes inside the greenhouse, on the effects of air leakage, and the influences of greenhouse shape and size, method of heating and crop arrangement. Another area in which inadequate data or understanding prevented realistic modelling was the role of the stomata in regulating gas exchange between the leaf and the greenhouse air.

These models aim at achieving a full description of the physical processes which take place in the greenhouse. An alternative type of model, the black box model, can be used if only a few unknown climatic factors are of interest as a function of a few known factors. It is then not necessary to consider the physical processes, as a transfer function between the output and the input

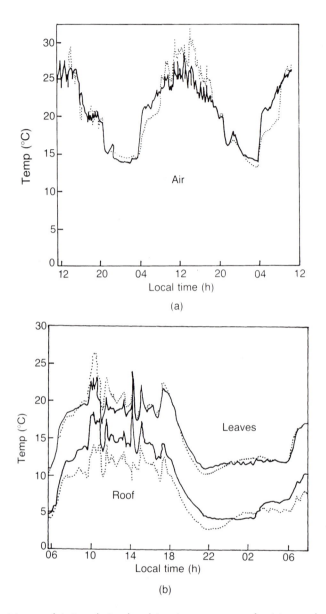

Fig. 7.2. Measured (—) and simulated (----) temperatures for (a) greenhouse air and (b) leaves and roof. (Bot, 1983.)

factors can be defined in such a way that the response of the output to variations in the input is represented in a proper manner. This form of model was used by Udink ten Cate (1983) to analyse the behaviour of piped heating systems, and by Hashimoto *et al.* (1981) to determine the response of plants to the greenhouse climate.

Advances in understanding the physical processes

The emphasis in dynamic modelling is on simulating the response of the greenhouse on a small timescale which requires the proper representation of the exchange processes between the interacting elements. It is no longer possible to assume, as with static models, that the exchange coefficients are constants and can be represented by the average values obtained by integration over long time periods. The heat and mass transfer coefficients are frequently functions of the system variables and it is important that they are formulated under conditions relevant to the greenhouse situation. As indicated by the comparison of the three time-dependent simulation models there are deficiencies in the understanding of a number of important physical processes. Recent improvements to the knowledge in some of these areas are outlined below.

Solar radiation transmission

An accurate description of irradiance in the greenhouse is important because solar radiation is by far the largest source of energy. The global radiation falling on a greenhouse consists of direct sun, which is assumed to include the circumsolar radiation, and diffuse sky radiation. Direct radiation can be predicted from a knowledge of the latitude, time of year and time of day. The amounts of direct and diffuse radiation transmitted into a greenhouse are calculated by tracing the path of a ray of light from a sky element to an element of the greenhouse cover, applying Fresnel's Law, which requires information on the optical properties of the cover, and the geometry and orientation of the greenhouse, and then following the paths of the resultant transmitted and reflected rays until either they are intercepted by the crop or greenhouse floor or leave the house. Repeating this for all sky elements and all elements of the greenhouse surface, including structural members, allows the radiation transmitted into the greenhouse to be calculated.

Kozai *et al.* (1978) developed such a model but did not include the reflection of light at the internal surfaces of the cover and considered only the incident beam. Critten (1983a,b) included reflections from the internal surfaces of the glazing material and concluded that they made a significant contribution to the overall greenhouse light transmissivity. The influence of the greenhouse structure in absorbing both direct and reflected light was investigated by Kurata (1989), who found the effect on reflected light was twice that on direct light.

An alternative way of treating the angular variation of global radiation based on harmonic analysis has been proposed by Critten (1986). The variation of radiance with the angles of azimuth and elevation is represented by the first few terms in a series expansion of the radiance distribution as orthogonal harmonic components using the product of the nth Legendre polynomial and the mth Fourier term to give the (n, m)th term of the distribution. The predicted variations in skylight distribution are shown in Fig. 7.3, and indicate considerable enhancement of the antisymmetric and second symmetric terms under direct sunlight conditions. Each harmonic component comprises a complete hemispherical distribution of known amplitude variation. When it passes through a transmitting surface, the amplitude will be modified and the resulting angular distribution can again be described in harmonic terms. In general, each incident harmonic term can contribute to all the harmonic terms in the transmitted distribution. To obtain the irradiance on a horizontal surface within the greenhouse, the problem reduces to determining just the zero-order harmonic in azimuth and the first-order harmonic in elevation as all other harmonics average to zero.

Information on the solar energy entering the greenhouse is required to establish its thermal status, but it is also required as an input to crop-growth models. In the former case gradients in solar irradiance within the greenhouse

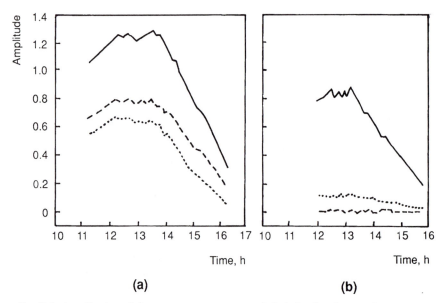

(a)   **(b)**

Fig. 7.3. Amplitudes of the major components of skylight distribution for (a) a stable blue sky and (b) a stable overcast sky. — coefficient of 1st symmetric component; ---- coefficient of 1st asymmetric component; ···· coefficient of 2nd symmetric component. (Critten and Stammers, 1986.)

are of little consequence, but this is not so for crop models. The ultimate object of crop-growth models is to predict the yield of the marketable product, which necessitates modelling photosynthesis, plant growth, and the development of fruit and flowers. To represent crop photosynthesis correctly the spatial distribution of photosynthetically active radiation (PAR), and the angular distribution, of the irradiation must be known because of the non-linear dependence of photosynthesis on light and because plants are frequently grown in rows.

Thermal radiation exchange

One of the less satisfactory features of current models is the treatment of long-wave radiation from the sky. In most cases semi-empirical relationships such as Brunt's are used in which the radiation is expressed as functions of air temperature and water vapour pressure measured in meteorological screens. This precludes a realistic treatment of partially cloudy skies. Atmospheric long-wave radiation can be measured directly using a modified net radiometer, but this instrument also measures global radiation which must therefore be deducted. Sky temperature could be used as a boundary condition if it could be measured directly (Seginer *et al.*, 1988).

Advances have been made by Silva and Rosa (1987) and Papadakis *et al.* (1987) in determining the exchange of thermal radiation between surfaces in the greenhouse. The latter authors presented a generalized theory for the exchange of thermal radiation between an arbitrary number of surfaces forming an enclosure, some of which can be partially transparent.

Heat and mass transfer coefficients

The exchanges of energy and mass occurring in the greenhouse system are characterized by heat and mass transfer coefficients. Many of the convective heat fluxes are calculated from empirical correlations based on the Nusselt, Grashof, Prandtl and Reynolds numbers, which are found in the heat transfer literature such as McAdams (1954). These relationships were frequently derived from data measured under idealized conditions and they are not necessarily directly applicable to the greenhouse. Bot (1983) has shown that the convective transfer of heat from the outer surface of a large multi-span greenhouse roof occurs in the transition region between free and forced convection (see Fig. 7.4). Kimball (1981) considered the influence of rain on the external coefficient and introduced a switching function. The coefficient of convective heat transfer at the inner surface of the greenhouse has been shown, by Kanthak (1970), Tantau and von Zabeltitz (1974) and Okada (1980), to be influenced by the type and position of the greenhouse heating system.

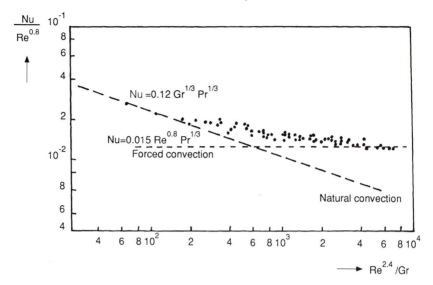

Fig. 7.4. Convective heat transfer at the outside of a multi-span glasshouse roof. Nu – Nussett number; Gr – Grashof number; Pr – Prandtl number; Re – Reynolds number. (Bot, 1983.)

A component of thermal models which assumes importance in summer is the prediction of ventilation. The major advances in greenhouse thermal models were stimulated by increases in the cost of energy which occurred in the mid-1970s. Ventilating greenhouses requires little energy so this process was not considered in detail in the models, other than to specify the rate of air exchange required to keep the greenhouse temperature at specified values. Baytorun (1986) showed that the rates of ventilation measured in a fan ventilated greenhouse agreed well on average with those predicted using a static energy balance, but there was considerable scatter of the data. Bot (1983) and de Jong (1990) correlated measured ventilation rates with the driving forces of wind speed and temperature difference between the inside and outside, and with the dimensions and angle of opening of the ventilators. Although the functional relationship of ventilation rate with these variables was correctly represented for their experimental greenhouses, ventilation rates were underestimated when the method was applied to other greenhouses. Okushima *et al.* (1989) used computational fluid dynamics to determine the flow of air into and through greenhouses during natural ventilation and the results agreed qualitatively with those obtained from studies with models in a wind tunnel (Sase *et al.*, 1984).

The response times of the elements in the greenhouse indicate that the soil is one component which should be analysed as a distributed factor. Most models use a one-dimensional treatment which is satisfactory when considering large greenhouses as the soil heat flux is only 5–15% of the total.

Chandra *et al.* (1981) considered the two-dimensional conduction of heat in the soil to determine the size of the edge effects. Papadakis *et al.* (1989) used an energy balance to investigate the thermal behaviour of the surface layer of soil. Soil heat flows are generally analysed using an apparent thermal conductivity which includes the effects of soil moisture on heat transfer, although Takakura *et al.* (1982) considered the three-dimensional flow of heat and water in an analysis of energy storage in the soil.

Transpiration

The process of transpiration makes crucial contributions to the energy balance of leaves and the moisture status of the greenhouse atmosphere. The evaporation rate has two driving forces: net radiation and the vapour pressure deficit of the air. To pass into the air from the leaf the vapour has to overcome two resistances: the stomatal resistance of the leaf and the resistance of the aerodynamic boundary layer. These features are described by the Penman–Monteith combination equation (Monteith, 1973). The satisfactory use of this relationship requires values for the two resistances. Values for stomatal resistance have recently been reported by Stanghellini (1987), Yang *et al.* (1990) and Jolliet and Bailey (1990). Stanghellini found that solar irradiance, and the temperature, vapour pressure deficit and CO_2 concentration of the air exerted the most significant influences on stomatal resistance. A transpiration model was proposed in which the influence of irradiance was represented as a rectangular hyperbola and parabolic functions used for the remaining factors. Use of this model, together with an energy balance to determine leaf temperature, enabled the observed transpiration of a tomato crop to be modelled closely (Fig. 7.5).

Sensitivity of models to parameter values

The more comprehensive greenhouse models contain a large number of coefficients and parameters whose values are often not known with great precision. The sensitivity of models to parameter values was investigated by Duncan *et al.* (1981) and Avissar and Mahrer (1982), who used models with a fixed set of input data and a range of parameter values. Chalabi and Bailey (1991) developed a method, based on the finite difference method, of determining the gradient of the state variables of the model with respect to the parameter value. The simulation model was executed using the Advanced Continuous Simulation Language (ACSL) and was used at least twice for each time step in order to determine the gradient. It was shown that the sensitivities were in general time-dependent and non-stationary. This method has the advantage that the sensitivity measures are explicitly defined and analytically exact because they are expressed in differential form. Also they were

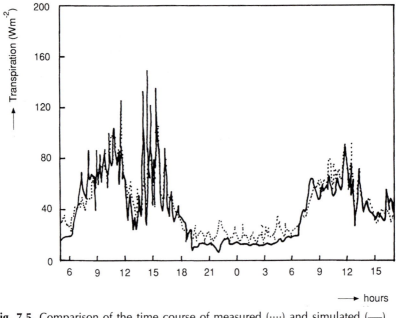

Fig. 7.5. Comparison of the time course of measured (····) and simulated (——) transpiration. (Stanghellini, 1987.)

normalized and dimensionless, which facilitated making comparisons, and the proportional change in a state variable could be determined from the sum of the proportional changes in the parameters multiplied by their respective sensitivities. This study shows that two parameters which had large influences on the temperature and humidity in the greenhouse were the solar absorptivity of the crop and the ratio of the greenhouse cover and ground areas.

Climate control

Controlling the greenhouse climate to produce a high-value crop has been considered as a multi-level control process (Udink ten Cate *et al.*, 1978) in which the levels are distinguished by their respective timescales (Fig. 7.6). At the lowest level is control of conditions in the greenhouse on a timescale of minutes. The middle level is concerned with control of plant development, where the timescales are set by physiological processes and can range from an hour to several days. The highest level of control deals with the planning, timing etc. of the overall production of the crop, and has a timescale which ranges from those in the middle level to the life of the crop. Control at the

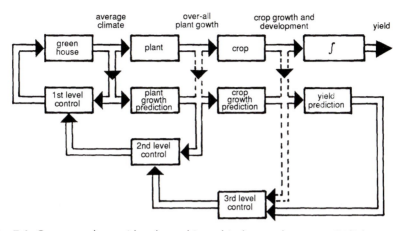

Fig. 7.6. Crop growth considered as a hierarchical control process. (Udink ten Cate, 1983.)

lowest level is implemented using feedback control, which has recently been supplemented with feedforward control obtained using greenhouse models, and system performance can be improved by adaptive control. At the middle level, control is based on various set-point values selected in the highest level, but which are modified by growers according to experience and appearance and performance of the crop. This, essentially open-loop, control could be converted to closed-loop control if plant growth was quantifiable; it would then be possible to modify the set-points used at the lowest level to improve the overall performance. Models of plant growth and production costs are beginning to be developed which can be used to optimize control at the lowest level (Seginer *et al.*, 1986). At the highest level, decisions are made in relation to the planning of the overall production of the crop. These are based on the rates of plant growth and development obtained in experiments to establish the crop-specific temperature blueprints. Recently, static regression models for planning production have been developed for a number of greenhouse crops including lettuce and cucumbers (Krug, 1988; Liebig, 1989). These were derived from experiments which used successive plantings and harvests during a season with different temperature set-points. Response surfaces were established which allowed growth rates and production periods to be predicted as a function of irradiance and temperature. These models enable decisions to be made on the consequences of changing the set-points of the controlled climatic variables in order to modify harvest dates, to reduce production costs or to alleviate the effects of variations in global radiation.

Discrete time control algorithms

When digital control systems began to be used for controlling the greenhouse climate, the control algorithms were generally a discrete version of those used in the analogue controllers. The normal method of heating control was via the heating pipe temperature which was regulated by a three-way valve that mixed return water from the greenhouse with hot water from the boiler. A master–slave configuration was used in which the master determined the required water temperature and the slave operated the control valve. The problem of tuning the proportional plus integral (PI) algorithm of the digital master controller, made difficult in greenhouse applications because of variations in the external conditions, was addressed by Udink ten Cate (1983). A relationship between pipe and air temperatures was obtained by approximating the heat flows in the greenhouse by a first-order transfer function with a dead time. Values for the parameters in the transfer function were estimated from experiments in which perturbations were made to the greenhouse set-point temperature at a frequency greater than the changes occurring in external temperature. The measured temperature signals were filtered and the parameters estimated by optimizing the fit between the calculated and observed times series in temperature. A disadvantage with discrete time PI control is integral action wind-up caused by saturation of the control signal due to constraints in the control loop. Udink ten Cate (1983) developed a modified PI algorithm to reduce the overshoot and undershoot in temperature which this produced following the night-to-day and day-to-night changes in set-point temperature. Udink ten Cate and van Zeeland (1981) designed a dog-lead PI algorithm, in which limits were placed on the controller output to bring it out of saturation sooner. Simulations (Udink ten Cate, 1983) demonstrated the superiority of the dog-lead algorithm, particularly in reducing undershoot, the most undesirable feature from the horticultural viewpoint (Fig. 7.7).

Davis and Hooper (1991) designed a controller which gave robust control in both normal and insulated greenhouses whose behaviours were of first

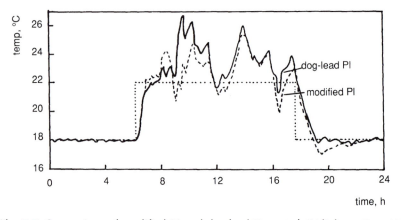

Fig. 7.7. Comparison of modified PI and dog-lead PI control. (Udink ten Cate, 1983.)

and second order respectively. The feedback signal was obtained from the difference between successive pairs of measured pipe and air temperatures and between the air temperature and its set-point. The gain coefficients were obtained from the measured temperature responses to pseudo-random input signals and the desired positions of the poles for the closed-loop system. Controller performance was improved by using a relatively long time sampling and control interval of 10 minutes, and it was found that filtering the measured air temperatures to reduce noise reduced valve movement without significantly degrading the accuracy of control.

Simulation models for control

One of the limitations of feedback control is that control actions cannot be taken until a change in greenhouse temperature has been detected, and control accuracy is further limited by the response time of the greenhouse to control actions. Computer-based control systems enabled both of these shortcomings to be addressed by the introduction of models to describe the dynamic behaviour of the greenhouse. These allowed the effects of changes in external conditions to be predicted, and thus corrective action could be taken earlier. When combined in the normal control loop the result became known as feedforward–feedback control (Fig. 7.8).

In principle, dynamic physical models of the greenhouse could be used in controller analysis and design. However, physical models have generally been used to simulate the response of the greenhouse to slowly varying input signals and thus the emphasis has been on low frequency response, whereas in control accurate high-frequency responses are important. Also the physical models are often complex and not suitable for the high-speed execution required in control applications. A control model of greenhouse air temperature was developed by Udink ten Cate (1983). This included the external weather factors of solar radiation, long-wave sky radiation, air temperature and wind speed, heating and ventilation. The model was idealized, as the greenhouse was considered as a perfectly stirred tank in which all variables had uniform values. The sensible heat fluxes were represented by a first-order differential equation which was linearized about the working air temperature. Laplace transformations yielded first-order transfer functions with parameters which could be identified with the ratios of the physical properties of the greenhouse and the heat transfer coefficients. Values for the transfer function parameters were obtained from experiments in which high-frequency perturbations were applied to the greenhouse heating and the temperature responses filtered before parameter estimation techniques were used. Further experiments gave the heat transfer coefficient of the heating pipes, which enabled the thermal properties to be obtained from the ratios. Udink ten Cate (1987) subsequently developed a more general model to

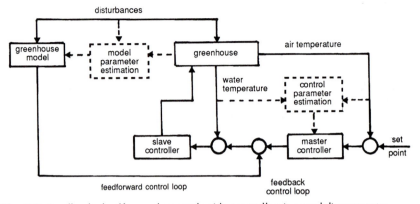

Fig. 7.8. Feedback–feedforward control with controller (or model) parameter adaption.

include the soil, which was treated as a single layer. These models were not formulated in terms of the position of the heating valve because of the non-linearity of valve response and the variability of valve characteristics, although some empirical relationships were proposed.

Adaptive control

In practice, the transfer coefficients of the greenhouse are not constants but vary with the external weather conditions; as a consequence the parameters of the transfer functions also vary. These parameter variations can be taken into account using adaptive control. Two approaches have been adopted: adaptive tuning of the control algorithm, and adaption of the control model.

Udink ten Cate and van de Vooren (1981) developed the first approach. They used an on-line, recursive parameter estimation algorithm to optimize the gains of the modified PI and dog-lead algorithms. A comparison of algorithm performance made using simulation showed that adaption gave only marginal benefits for both algorithms. There are a number of problems associated with applying this technique to real processes. The parameters for adaption have to be identified in advance. Satisfactory parameter estimation requires reasonably large variations in the inputs in order to excite the system sufficiently, and these may not occur. Over long time periods there is the possibility that conditions occur which cause the estimation process to produce extreme parameter values.

The second approach was taken by Tantau (1985). A basic static energy balance model was used to relate the heat supplied by the heating system and by solar radiation to the heat lost from the greenhouse. This enabled the heating set-point temperature required to achieve a given internal air temperature to be defined in terms of the internal and external temperatures, solar irradiance and the heat transfer coefficients of the heating pipes and the greenhouse. Two parameters were defined which contained the heat transfer coefficients and factors related to the conversion of external global radiation to internal sensible heat. Because the transfer coefficients depended on the physical conditions these parameters were not constant and had to be estimated on-line. This was achieved using relationships, derived from the model, involving the parameters, solar radiation, and the temperatures of the heating pipes and internal and external air, which were all measured by the control system. The resulting control temperature was stable and close to the set-point (Fig. 7.9). However, difficulties were encountered if the heat transfer coefficients changed sharply, as occurs when a thermal screen was closed or opened, as the parameter adaption took a long time. Consequently it was suggested that the parameter values should be switched by the control algorithm so the adaption was only required to follow relatively slow rates of change.

On-line statistical time series techniques were used by Davis (1984) to provide adaptive control of greenhouse ventilation. A linear time series formula with variable parameters was fitted to successive samples of ventilator aperture and greenhouse temperature and used to predict the ventilator adjustment necessary to maintain the desired temperature. Simulation studies made using a physical model of the non-linear relationship between internal temperature and ventilator aperture showed that adaptive control was superior to PI control with fixed parameters. Adaptive control using a physical model of greenhouse ventilation was used by Tantau (1989), in which three parameters were adapted on-line.

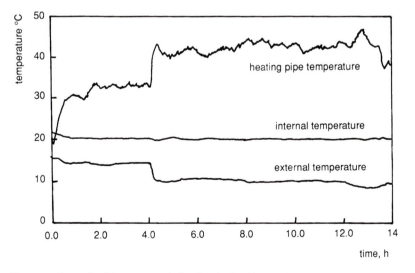

Fig. 7.9. Control of heating with feedback–feedforward control using greenhouse model with parameter adaption. (Tantau, personal communication.)

Multivariable control

The greenhouse climate is governed by a multivariable process with several inputs and outputs and strongly coupled state variables. The classical approach to controlling such a system has been to control each variable by a single-loop controller and ignore the coupling. The multivariable nature of the greenhouse climate is described by current greenhouse climate models, although they do not give a true description of the real control behaviour of the greenhouse system because dead times and actuator characteristics are not included. However, these models can be used to design multivariable controllers which expressly allow for coupling of the variables. Van Henten (1989) used simulations to show that feedback of the additional information made possible with multivariable control improved controller performance: control accuracy was improved, and there was less overshoot and no saturation of the actuator signals. Franceschi *et al.* (1989) demonstrated that greenhouse temperature and humidity could be controlled satisfactorily by a multivariable controller.

The 'speaking plant' approach to control

Adequately modelling the influence of the greenhouse climate on the physiological activity of plants represents a formidable problem because of the complexity of both the physical and physiological components. The 'speaking plant' approach to control seeks to overcome this problem by measuring the plant responses directly (Hashimoto *et al.*, 1981). The plant is considered as a black box model, and system identification techniques are used to determine the responses of plants to changes in climatic variables. Responses in the time domain were identified by monitoring the changes following a step change in plant irradiance. Responses in the frequency domain were determined by using a random binary signal to change the irradiance and applying spectral analysis to the output signals. This information enabled impulse responses and, subsequently, step responses to be calculated which matched the observed behaviour (Hashimoto *et al.*, 1982). Using these techniques, Hashimoto (1982) identified changes in leaf temperature resulting from changes in stomatal aperture caused by the onset of water stress. Leaf temperature was then used to control irrigation to maintain maximal stomatal aperture to maximize CO_2 uptake and consequently photosynthesis. The influences of irradiance and CO_2 concentration on photosynthesis were studied using a linear model (Hashimoto *et al.*, 1985) and a more realistic non-linear model (Hashimoto and Yi, 1989) of photosynthesis.

As a result of this work, Hashimoto *et al.* (1985) proposed that two types of plant response should be used in greenhouse climate control: one obtained from system identification, and the other obtained by monitoring the short-term plant behaviour such as stomatal conductance, net CO_2 uptake and leaf temperature. The former responses would form a database in the climate control computer and be used in conjunction with the latter to determine the most suitable control option.

Optimization

The ultimate objective of greenhouse climate control is to manipulate the climate in a way which maximizes the difference between the value of the crop and the costs of production (Fig. 7.10). To achieve this completely would require optimal control being exercised at each level of the multi-level control scheme. Although this is not possible at present, it is practicable to consider optimal control for some climatic variables and the timing of certain control actions taken at the lowest level. A prerequisite is the availability of suitable greenhouse, crop and economic models to quantify the various factors.

In the case of temperature control over short periods there is evidence that the rate of plant growth is closely related to the average temperature and that, within limits, deviations in temperature have no influence (Hurd and Graves, 1984). As a consequence, Bailey (1985) proposed the greenhouse temperature should be lowered during windy conditions and raised under calm conditions to achieve the required temperature integral, but with a reduced energy input. This should not influence crop productivity but will reduce production costs. Another application where a simple plant model can be used is in the control of thermal screens to reduce greenhouse heat loss. Bailey (1988) considered the loss of production was proportional to the light integral, and established the optimum screen operating condition at which the value of the energy saved by closing the screen equalled the loss in crop revenue resulting from the reduced light.

A more detailed plant model was used by Challa and van de Vooren (1980) to give an economic value to the earliness of cucumber production, and to calculate optimal greenhouse temperatures for a range of light conditions with an emphasis on the rate of leaf initiation and growth. Considering the rate of carbon assimilation by a mature cucumber crop under steady state conditions, Challa and Schapendonk (1986) related the value of the additional dry matter produced to the cost of carbon dioxide enrichment. The optimal CO_2 enrichment strategy for tomato seedling production which minimized the combined costs of CO_2 enrichment and heating was determined by Seginer *et al.* (1986). Analytical relationships for the optimal CO_2 concentration for lettuce were established as a function of irradiance, ventilation rate, and the ratio of lettuce price to CO_2 cost, by Chalabi and Critten (1990), and have been developed into a form suitable for maximizing grower income.

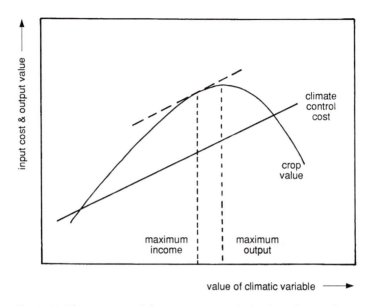

Fig. 7.10. The concept of the economic optimization of control.

The formal approach to optimal control developed by Seginer was based on Pontryagin's Maximum Principle. This permits the problem of finding the optimal control sequence which maximizes the chosen objective function over a complete season to be transformed to one of continuously maximizing a function called the Hamiltonian. This represents the sum of the current net gains and those which will be obtained in the future as a consequence of control decisions made now. The crucial parameter in this theory is the adjoint variable, which is the value attached to a marginally produced unit of output. The best value of the adjoint variable is established by simulations to determine the optimum solution. This method was applied to the control of greenhouse heating and thermal screens by Bailey and Seginer (1989). It was found the adjoint variable was only piece-wise constant, so the control was sub-optimal. An alternative method is the Simplified Control Search (SCS) used by Marsh *et al.* (1987) to select the best daytime temperature set-points for lettuce. At the time of each control decision, simulations over the remainder of the season are used to find the one which is most profitable. Starting with a reasonable control policy, such as a constant set-point, SCS selects at each decision time a decision which is as good as or better than the original, because all feasible policies, including the original one, are assessed.

Artificial intelligence

The advances in climate control have occurred largely as a result of improved quantitative descriptions of the greenhouse climate, control system behaviour or the response of plants to the climate. This deterministic approach requires a detailed understanding of the system before progress can be made. An alternative, which does not require the system to be fully defined, is to make use of artificial intelligence.

Hoshi and Kozai (1984) designed a knowledge-based control system in which the climate control strategies were expressed as production rules (if . . . then rules). These were contained in a knowledge base which was independent of the control program in order to facilitate the introduction of new knowledge, which could be provided by the grower. An operational interface with the grower was provided using inference, explanation and natural language programs to explain the reasons for selecting the control strategy being implemented at any time.

A number of ways of capturing the knowledge and experience of expert growers have been explored in order to develop expert systems which can be used in control applications. Takakura *et al.* (1984) developed a consultation system for the management of tomato crops, with particular reference to irrigation, based on records of the management decisions made by expert growers. Routine monitoring of set-points, control actions, and climatic conditions inside and outside a number of commercial production greenhouses was used by Hirosawa and Kozai (1987) to establish a knowledge base on how growers manage greenhouses. Kurata (1988) developed an algorithm which determined the production rules by analysing such a database. In contrast to this automated way of establishing the knowledge base, Jacobson *et al.* (1989) elicited information by interviews with an expert to obtain a knowledge base on the use of misting during plant propagation. This was combined with a real-time monitoring and control system to implement the expert's perceived optimal misting strategy. An expert system was used by Jones and Haldeman (1986) to diagnose failures in the environmental control systems of growth chambers, and was of benefit even to experienced operators.

Harazono *et al.* (1984) used a computerized learning method as an aid to maximizing the rate of photosynthesis of young cucumber plants. An initial database of photosynthesis and transpiration rates was established by experiments covering a range of irradiances, temperatures, vapour pressure deficits and carbon dioxide concentrations. During control, the database was interrogated using the measured irradiance to find the values for the other variables which gave the maximum photosynthetic rate. The database was improved by the addition of information obtained during control to give a progressively more accurate representation of the plant response.

The grower–controller interface

The increasing accuracy and reliability of physical models of the greenhouse coupled with the advances being made in understanding plant response and growth is leading towards a more rational control of the greenhouse climate over the short term. However, there are some potential difficulties. One is that the models, particularly the biological ones, are unlikely to describe all the relevant reactions of the crop to its environment. Another is that although the models are tested by comparing their predictions with observation, the validation process is unlikely to cover the extremes of conditions which occur in nature. A third problem area is that the plant models generally relate to plants which are not subjected to stress, but in practice it is questionable whether, even in greenhouses, plant stress can always be prevented. These limitations, coupled with the fact that a crop consists of a community of plants which differ individually because of natural variability, means it will be necessary for the grower to make the overall control decisions. The use of expert systems will assist the grower in reaching decisions on the basis of his or her experience and observations of the crop, and the historic data provided by the controller. A requirement for controllers implementing optimal control will be an explanation facility to explain the logic behind the selection of particular set-points and to demonstrate that the controller is functioning correctly. Thus an intelligent interface will form an important part of the overall control system and will play a vital role in the acceptance of the new generation of control systems by the growers.

References

Albright, L. D., Seginer, I., Marsh, L. S. and Oko, A. (1985) In situ thermal calibration of unheated greenhouses. *Journal of Agricultural Engineering Research* 31, 265–81.

Avissar, R. and Mahrer, Y. (1982) Verification study of a numerical greenhouse microclimate model. *Transactions of the American Society of Agricultural Engineers* 25, 1711–20.

Bailey, B. J. (1981) The reduction of thermal radiation in glasshouses by thermal screens. *Journal of Agricultural Engineering Research* 26, 215–24.

Bailey, B. J. (1984) Limiting the relative humidity in insulated greenhouses at night. *Acta Horticulturae* 148, 411–19.

Bailey, B. J. (1985) Wind dependent control of greenhouse temperature. *Acta Horticulturae* 174, 381–6.

Bailey, B. J. (1988) Control strategies to enhance the performance of greenhouse thermal screens. *Journal of Agricultural Engineering Research* 40, 187–98.

Bailey, B. J. and Seginer, I. (1989) Optimum control of greenhouse heating. *Acta*

Horticulturae 245, 512–18.

Baille, A., Aries, F., Baille, M. and Laury, J. C. (1985) Influence of thermal screen properties on heat losses and microclimate of greenhouses. *Acta Horticulturae* 174, 111–18.

Baytorun, N. (1986) *Bestimmung des Luftweschsels bei geluften Gewachshausern.* Gartenbautechnische Informationen, Heft 27. Institut für Technik in Gartenbau und Landwirtschaft, University of Hannover.

Bot, G. P. A. (1983) 'Greenhouse climate: from physical processes to a dynamic model.' Ph.D. Thesis, Agricultural University, Wageningen, The Netherlands.

Businger, J. A. (1963) The glasshouse (greenhouse) climate. In: W. R. Wijk (ed.), *Physics of Plant Environment.* North Holland Publishing Co., Amsterdam, 277–318.

Chalabi, Z. S. and Critten, D. L. (1990) The influence of real time filtering strategies on optimal CO_2 concentration for a greenhouse lettuce crop. *Acta Horticulturae* 268, 139–48.

Chalabi, Z. S. and Bailey, B. J. (1991) Sensitivity analysis of a non-steady state model of the greenhouse microclimate. *Agricultural and Forest Meteorology* 56, 111–27.

Challa, H. and van de Vooren, J. (1980) A strategy for climate control in greenhouses in early winter production. *Acta Horticulturae* 106, 159–64.

Challa, H. and Schapendonk, A. H. C. M. (1986) Dynamic optimization of CO_2 concentration in relation to climate control in greenhouses. In H. Z. Enoch and B. A. Kimball (eds), *Carbon Dioxide Enrichment of Greenhouse Crops*, vol 1. CRC Press Inc., USA, 146–60.

Chandra, P., Albright, L. D. and Scott, N. R. (1981) A time dependent analysis of greenhouse thermal environment. *Transactions of the American Society of Agricultural Engineers* 24, 442–9.

Cooper, P. I. and Fuller, R. J. (1983) A transient model of the interaction between crop, environment and greenhouse structure for predicting crop yield and energy consumption. *Journal of Agricultural Engineering Research* 28, 401–17.

Critten, D. L. (1983a) A computer model to calculate the daily light integral and transmissivity of a greenhouse. *Journal of Agricultural Engineering Research* 28, 61–76.

Critten, D. L. (1983b) The evaluation of a computer model to calculate the daily light integral and transmissivity of a greenhouse. *Journal of Agricultural Engineering Research* 28, 545–63.

Critten, D. L. (1986) A general analysis of light transmission in greenhouses. *Journal of Agricultural Engineering Research* 33, 289–302.

Critten, D. L. and Stammers, V. A. (1986) The use of natural skylight variations to measure the light transmissivity of harmonic components of skylight in greenhouses. Divisional Note DN 1350, National Institute of Agricultural Engineering, Silsoe, Beds, UK.

Davis, P. F. (1984) A technique of adaptive control of the temperature in a greenhouse using ventilator adjustments. *Journal of Agricultural Engineering Research* 29, 241–8.

Davis, P. F. and Hooper, A. W. (1991) Improvement of greenhouse heating control. *Proceedings of the Institution of Electrical Engineers. Part D: Control Theory and Applications* 138(3), 249–55.

de Jong, T. (1990)'Natural ventilation of large multi-span greenhouses.' Ph.D. Thesis, Agricultural University, Wageningen, The Netherlands.

Deltour, J., de Halleux, D., Nijskens, J., Coutisse, S. and Nisen, A. (1985) Dynamic modelling of heat and mass transfer in greenhouses. *Acta Horticulturae* 174, 119–28.

den Hartog, F. (1988) 'Numerical simulation of the turbulent air flow and heat transfer in unheated and heated greenhouses.' M.Sc. Thesis, Technical University of Delft, The Netherlands.

Duncan, G. A., Loewer, O. J. and Colliver, D. G. (1981) Simulation of energy flows in a greenhouse: magnitudes and conservation potential. *Transactions of the American Society of Agricultural Engineers* 24, 1014–21.

Franceschi, M., Oueslati, L. and Enea, G. (1989) Regulation multivariable d'une serre agricole. In: J. P. Sagaspe and A. Villeger (eds), *Agrotique 89. Proceedings of the Second International Conference, Bordeaux, France, 26–28 September 1989.* TEKAEA, France, 141–53.

Froehlich, D. P., Albright, L. D., Scott, N. R. and Chandra, P. (1979) Steady-periodic analysis of glasshouse thermal environment. *Transactions of the American Society of Agricultural Engineers* 22, 387–99.

Garzoli, K. V. and Blackwell, J. (1981) An analysis of the nocturnal heat loss from a single skin plastic greenhouse. *Journal of Agricultural Engineering Research* 26, 203–14.

Garzoli, K. V. and Blackwell, J. (1987) An analysis of the nocturnal heat loss from a double skin plastic greenhouse. *Journal of Agricultural Engineering Research* 36, 75–85.

Harazono, Y., Taenaka, T. and Yabuki, K. (1984) Optimising environmental control of greenhouse climate for cucumber growth by means of 'learning control method'. *Acta Horticulturae* 148, 259–65.

Hashimoto, Y. (1982) Dynamic behaviour of leaf temperature: a review. *Biological Science (Tokyo)* 34, 68–75.

Hashimoto, Y. and Yi, Y. (1989) Dynamic model of CO_2 uptake based on system identification. *Acta Horticulturae* 248, 295–300.

Hashimoto, Y., Morimoto, T. and Funada, S. (1981) Computer processing of speaking plant for climate control and computer aided plantation (computer aided cultivation). *Acta Horticulturae* 115, 317–25.

Hashimoto, Y., Morimoto, T. and Funada, S. (1982) Identification of water deficiency and photosynthesis in short-term plant growth under random variation of the environment. In: *Sixth IFAC Symposium on Identification and System Performance Estimation*, vol. 2. Arlington, USA.

Hashimoto, Y., Morimoto, T. and Fukuyama, T. (1985) Some speaking plant approaches to the synthesis of control system in the greenhouse. *Acta Horticulturae* 174, 219–26.

Hirosawa, Y. and Kozai, T. (1987) Remote monitoring/control system for greenhouse computers. *Nogyo oyobi Engei* 62(4), 553–8.

Hoshi, T. and Kozai, K. (1984) Knowledge-based and hierarchically distributed online control system for greenhouse management. *Acta Horticulturae* 148, 301–8.

Hurd, R. G. and Graves, C. J. (1984) The influence of different temperature patterns having the same integral on the earliness and yield of tomatoes. *Acta Horticulturae* 148, 547–54.

Jacobson, B. K., Jones, P. H., Jones, J. W. and Paramore, J. A. (1989) Real-time monitoring and control with an expert system. *Computers and Electronics in Agriculture* 3, 273–85.

Jolliet, O. J. and Bailey, B. J. (1990) Transpiration of greenhouse tomato plants as a function of airspeed, solar radiation, vapour pressure deficit and temperature. *Proceedings XXIII International Horticultural Congress.* International Society for Horticultural Science, Florence, Italy.

Jolliet, O. J., Gay, J.-B., Bourgeois, M., Danloy, L., Bretton, T., Mantilleri, S., Reist, A. and Moncousi, C. (1985) Solar gains and thermal rejects by ventilation. *Acta Horticulturae* 174, 127–34.

Jones, P. and Haldeman, J. (1986) Management of a crop research facility with a microcomputer-based expert system. *Transactions of the American Society of Agricultural Engineers* 29, 235–42.

Kanthak, P. (1970) Der Einfluss von Heizungssystemen mit unterschiedlichem Strahlungsanteil auf das Klima und den Warmehaushalt von Hallenbauten mit grossen Glasflachen, speziell von Gewachshausern. *Fortschritt-Berichte der VDI Zeitschriften,* series 6, no. 28. VDI-Verlag GmbH, Dusseldorf.

Kimball, B. A. (1973) Simulation of the energy balance of a greenhouse. *Agricultural Meteorology* 11, 243–60.

Kimball, B. A. (1981) A versatile model for simulating many types of solar greenhouses. *American Society of Agricultural Engineers,* paper no. 81-4038.

Kozai, T., Goudriaan, J. and Kimura, M. (1978) *Light Transmission and Photosynthesis in Greenhouses.* Wageningen Centre for Agricultural Publishing and Documentation, The Netherlands.

Krug, H. (1988) Basic principles of economic use of heat and CO_2 in greenhouse production. *Acta Horticulturae* 229, 265–72.

Kurata, K. (1988) Greenhouse control by machine learning. *Acta Horticulturae* 230, 195–200.

Kurata, K. (1989) Model of light environment in greenhouses with emphasis on the role of reflection. *Acta Horticulturae* 240, 109–14.

Liebig, H.-P. (1989) Model of cucumber growth and prediction of yields. *Acta Horticulturae* 248, 187–91.

McAdams, W. H. (1954) *Heat Transmission,* 3rd edn. McGraw-Hill, New York.

Marsh, L. S., Albright, L. D., Langhans, R. W. and McCulloch, C. E. (1987) Economically optimum day temperatures for greenhouse hydroponic lettuce production. *American Society of Agricultural Engineers* paper no. 87-4023.

Monteith, J. L. (1973) *Principles of Environmental Physics.* Edward Arnold, London.

Morris, L. G. and Winspear, K. W. (1967) The control of temperature and humidity in glasshouses by heating and ventilation. Paper 10 in: J. A. C. Gibb (ed.), *Proceedings of Agricultural Engineering Symposium,* Silsoe. Business Books Ltd, London, 1969; for the Institution of Agricultural Engineers.

Nara, M. (1979) Studies on the air distribution in farm buildings (1): two dimensional numerical analysis and experiment. *Journal of the Society of Agricultural Structures of Japan* 9, 18–25.

O'Flaherty, T., Gaffney, B. J. and Walsh, J. A. (1973) Analysis of the temperature control characteristics of heated greenhouses using an analogue computer. *Journal of Agricultural Engineering Research* 18, 117–32.

Okada, M. (1980) The heating load of greenhouse: (1) Convective heat transfer coefficients at the inside cover surface of a greenhouse as influenced by heating pipe positions. *Journal of Agricultural Meteorology* 35, 235–42.

Okada, M. (1985) An analysis of thermal screen effects on greenhouse environment by means of a multi-layer screen model. *Journal of Agricultural Meteorology*, 39, 103–6.

Okushima, L., Sase, S. and Nara, M. (1989) A support system for natural ventilation design of greenhouses based on computational aerodynamics. *Acta Horticulturae* 248, 129–36.

Papadakis, P., Frangoudakis, A. and Kyritsis, S. (1987) Theoretical and experimental investigation of thermal radiation transfer in polyethylene covered greenhouses. *Journal of Agricultural Engineering Research* 44, 97–111.

Papadakis, P., Frangoudakis, A. and Kyritsis, S. (1989) Soil energy balance analysis for a solar greenhouse. *Journal of Agricultural Engineering Research* 43, 231–43.

Sase, S., Takakura, T. and Nara, M. (1984) Wind tunnel testing on airflow and temperature distribution of a naturally ventilated greenhouse. *Acta Horticulturae* 148, 329–36.

Seginer, I. and Kantz, D. (1986) In-situ determination of transfer coefficients for heat and water vapour in a small greenhouse. *Journal of Agricultural Engineering Research* 35, 39–54.

Seginer, I. and Kantz, D. (1989) Night-time use of dehumidifiers in greenhouses: an analysis. *Journal of Agricultural Engineering Research* 44, 141–58.

Seginer, I., Angel, A., Gal, S. and Kantz, D. (1986) Optimal CO_2 enrichment strategy for greenhouses. A simulation study. *Journal of Agricultural Engineering Research* 34, 285–304.

Seginer, I., Kantz, D., Peiper, U. M. and Levav, N. (1988) Transfer coefficients of several polyethylene greenhouse covers. *Journal of Agricultural Engineering Research* 39, 19–37.

Short, T. H. and Breuer, J. J. G. (1985) Greenhouse energy demand comparisons for The Netherlands and Ohio, USA. *Acta Horticulturae* 174, 145–53.

Silva, A. M. and Rosa, R. (1987) Radiative heat loss inside a greenhouse. *Journal of Agricultural Engineering Research* 37, 155–62.

Soribe, F. I. and Curry, R. B. (1973) Simulation of lettuce growth in an air-supported plastic greenhouse. *Journal of Agricultural Engineering Research* 18, 133–40.

Stanghellini, C. (1987) 'Transpiration of greenhouse crops. An aid to climate management.' Ph.D. Dissertation, Agricultural University, Wageningen. The Netherlands.

Takakura, T. (1967 and 1968) Predicting air temperatures in the greenhouse, I and II. *Journal of the Meteorology Society of Japan* 45, 40–52; 46, 36–43.

Takakura, T., Jordan, K. A. and Boyd, L. L. (1971) Dynamic simulation of plant growth and the environment in the greenhouse. *Transactions of the American Society of Agricultural Engineers* 14, 964–71.

Takakura, T., Nishina, H. and Kurata, K. (1982) A simulation analysis of solar greenhouses with underground heat storage units. *Energy Conservation and Use of Renewable Energies in the Bio-Industries* 2, 634–9.

Takakura, T., Shono, H. and Hojo, T. (1984) Crop management by intelligent computer systems. *Acta Horticulturae* 148, 317–18.

Tantau, H.-J. (1975) *Der Einfluss von Einfach- und Doppelbedachungen auf das Klima und den Warmehaushalt von Gewachshausern.* Gartenbautechnische Informationen, Heft 4. Institut für Technik in Gartenbau und Landwirtschaft, University of Hannover.

Tantau, H.-J. (1984) The ITG digital greenhouse climate control system for energy saving. *Gartenbauwissenschaft* 49(3), 140-3.

Tantau, H.-J. (1985) Greenhouse climate control using mathematical models. *Acta Horticulturae* 174, 449–59.

Tantau, H.-J. (1989) Models for greenhouse climate control. *Acta Horticulturae* 245, 397–404.

Tantau, H.-J. and von Zabeltitz, Chr, (1974) Greenhouse heat requirement depending on different heating systems. *Proceedings XIX International Horticultural Congress.* International Society for Horticultural Science, Warsaw.

Udink ten Cate, A. J. (1983) 'Modelling and (adaptive) control of greenhouse climates.' Ph.D. Thesis, Agricultural University, Wageningen, The Netherlands.

Udink ten Cate, A. J. (1987) Analysis and synthesis of greenhouse climate controllers. In: Clark, J. A., Gregson, K. and Saffell, R. A. (eds), *Computer Applications in Agricultural Environments.* Butterworths, London.

Udink ten Cate, A. J. and van de Vooren, J. (1981) Adaptive systems in greenhouse climate control. In: *Preprints 8th IFAC World Congress*, Kyoto, Japan. Late papers, 9–15.

Udink ten Cate, A. J. and van Zeeland, J. (1981) A modified PI-algorithm for a glasshouse heating system. *Acta Horticulturae* 115, 351–8.

Udink ten Cate, A. J., Bot, G. P. A. and van Dixhoorn, J. J. (1978) Computer control of greenhouse climates. *Acta Horticulturae* 87, 265–72.

van Bavel, C. H. M., Damagnez, J. and Sadler, E. J. (1981) The fluid-roof solar greenhouse: energy budget analysis by simulation. *Agricultural Meteorology* 23, 61–76.

van Bavel, C. P. H., Takakura, T. and Bot, G. P. A. (1985) Global comparison of three greenhouse climate models. *Acta Horticulturae* 174, 21–33.

van Henten, E. J. (1989) Model based design of optimal multivariable climate control systems. *Acta Horticulturae* 248, 301–6.

von Elsner, B. (1982) *Das Kleinklima und der Warmeverbrauch von geschlossenen Gewachshausern.* Gartenbautechnische Informationen, Heft 12. Institut für Technik in Gartenbau und Landwirtschaft. University of Hannover.

Walker, J. N. (1965) Predicting temperatures in ventilated greenhouses. *Transactions of the American Society of Agricultural Engineers* 8, 445–8.

Whittle, R. M. and Lawrence, W. J. C. (1959 and 1960) The climatology of glasshouses: I, II, III, IV and V. *Journal of Agricultural Engineering Research* 4, 326–40; 5, 36–41; 5, 165–78; 5, 235–40; 5, 339–405.

Yang, X., Short, T. H., Fox, R. D. and Bauerle, W. L. (1990) Transpiration, leaf temperature and stomatal resistance of a greenhouse cucumber crop. *Agricultural and Forest Meteorology* 51, 197–209.

Chapter 8

Heat and Mass Transfer in the Drying and Cooling of Crops

J.L. Woods

Introduction

The production of agricultural commodities is seasonal, whereas the food industry requires a continuous supply. This gives rise to a sophisticated storage and distribution system. The quality is maintained in storage largely through the control of physical environment. Through the lowering of temperature and water activity, the biological activity of both the stored material and potential pests is minimized. In high-moisture materials, such as vegetables, fruits and meats, the moisture is fundamental to the structure and needs to be retained. Storage must then be achieved through lowering temperature, which can be combined with control of gaseous environment. In dry materials, such as grains, low water activity combined with secondary control over temperature enables long-term storage. It is clear, therefore, that the processes of drying and cooling are fundamental to the upstream management of the inputs to the food industry.

During the mass transfer process of drying, there is a simultaneous heat transfer process required to supply the latent heat of vaporization. In the cooling of biological materials, there will also be an associated moisture loss. In both cases it is necessary to describe the simultaneous processes of heat and mass transfer. These processes are illustrated in Fig. 8.1 for the cases of high- and low-moisture crops. They show drying and cooling in an air flow, which is the most commonly used approach and the subject of interest in this chapter. The data review and analysis presented here is primarily concerned with the convection processes in the air flow through packed beds. As will be discussed, in order to predict crop temperature, the convective heat transfer coefficient must be specified.

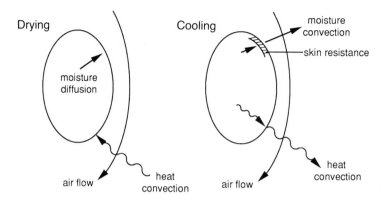

Fig. 8.1. Limiting transfer processes in the drying of low moisture and cooling of high moisture crops.

The limiting stages of the transfer processes in the drying of low moisture materials are illustrated in Fig. 8.1. Moisture transfer is limited by the diffusion rate within the particle, whilst the inward flow of heat is limited by the convective exchange across the boundary layer. At the high initial rates of moisture transfer that occur at higher air temperatures, the drying material must be substantially cooler than the air in order to transfer heat for evaporation. The external convective heat transfer coefficient is therefore critical in controlling material temperature. The internal moisture diffusion is itself sensitive to temperature and the heat and mass transfer processes are therefore very interactive.

The realization of this temperature differential effect has led to major developments in the design and operation of grain driers. The most effective way of taking advantage of this temperature differential is through the concurrent flow drier geometry, in which the hottest air meets the wettest grain, maximizing differential temperature. The exponential increase in moisture-carrying capacity of air with temperature reduces air flow requirements and makes such driers very compact and energy efficient (Muhlbauer et al., 1971; Isaacs and Muhlbauer, 1975).

A number of these driers were developed in the United States during the 1970s (Baughman, 1971; Anderson, 1972; Westelaken, 1977; Hall and Anderson, 1980) and inlet air temperatures as high as 500 °C were employed without grain damage. However, they have made no impact on the European market, where the mixed flow drier has become widely used, particularly at centralized drying locations. This has an element of concurrent flow and has gone some way to increasing the permissible air temperatures and hence performance, whilst retaining relative simplicity in design and operation (Anon, 1983a). With the use of recirculation strategies, the energy efficiency does not fall far below more sophisticated designs. A review of these developments was presented by Nellist (1982). Since increased air temperature reduces fan and containment costs, CO_2 output, energy consumption and air pollution, it is likely to remain a major objective in future drier development.

The grain-to-air temperature differential, upon which all these developments are based, is controlled by the convective heat transfer coefficient in the air boundary layer. This temperature differential is in fact related to both the evaporation rate and heat transfer coefficient. Although considerable data are available on evaporation rate, data on heat transfer coefficient for grains are limited. It is the objective here to draw these data together with research from other technologies, providing a basis for future developments in high-temperature drying of grains and other particulate materials.

In the design of driers, computer simulation must play a major role, due to the cost of experimentation on this scale and the difficulty of humidity and grain temperature measurement. In the development of control systems to handle the fluctuations of input grain moisture content, it is difficult to imagine the development of control algorithms without simulation. Future objectives of designers might be summarized as.

1. Control of output moisture content for fluctuating input values.
2. Maximization of input air temperature with grain temperature controlled just below the critical damage level.

To achieve this through simulation will require an improved formulation for the heat transfer coefficient.

An essential part of any drier is the cooling section. Current designs achieve cooling in sections of the same geometry as those used for drying. Given the totally different mechanisms controlling heat and moisture transfer, there is scope for the reassessment of grain cooling. The relationship between heat transfer coefficient, air velocity and grain diameter would be essential to such a study.

In the cooling of high moisture crops (Fig. 8.1), it is the external convective heat transfer coefficient that again tends to limit the flow of sensible heat. For large diameters and high air velocities the internal conduction resistance can be significant (Woods, 1990). Moisture transfer within these high moisture crops is generally not considered limited by internal diffusion. Resistance to moisture loss lies largely in the surface layer. The variation in skin resistance is very large. The potato tuber, for example, loses moisture at around 1/60th of the rate of the carrot or leafy vegetables. Where moisture loss is rapid, the external convective mass transfer coefficient has a significant effect.

There has been limited research on deep bed cooling of high-moisture crops as compared with the drying of grains. Their susceptibility to handling damage means that they cannot be handled as a flowing granular material. This has led to less standardized designs of cooling system based on containers such as pallet boxes. These systems are incorporated into structures and not sold as free-standing items of equipment. This, together with the diversity of crops processed, may account for a less unified approach to the cooling of high-moisture crops. The heat transfer analysis presented here will assist in developments in this area.

The work described in this chapter gathers together previous data and, with additional analysis, presents working correlating equations for particle-to-air convective heat and mass transfer coefficients in packed beds. Data on grains and from the wider field of chemical engineering are drawn together and compared. From this comparison an interesting insight into thermal dispersion in the flow direction is also obtained.

Heat transfer coefficient

In reviewing the experimental data on grain-to-air heat transfer coefficient in a packed bed, two sources of data are considered. Measurements made specifically on grain are presented first in the form of the volumetric heat transfer coefficient. The developments in experimental method and the problems of simultaneous moisture transfer are discussed in relation to grain drying. General packed bed data are then presented in dimensionless form for comparison with selected measurements on grains. The objective is to relate the specific grain measurements to the general data and thereby identify a correlating equation appropriate to grain drying simulation. Also, the upper Reynolds number limit on the correlations is examined to assess their suitability for use in the description of heat and moisture convection in the cooling of high-moisture crops, where larger diameters and velocities occur.

Data for grain beds

All the experiments reviewed employ some form of transient method and attempt to minimize the effect of moisture transfer. This is achieved by bringing the grain to equilibrium with the hot air stream and then cooling the sample in a sealed container prior to the heat transfer experiment. They can be subdivided into thin layer or deep bed techniques, but are discussed here in chronological order. The data are presented by each author in the form of correlating equations of the form:

$$h_v = e\,G^f \qquad\qquad [8.1]$$

where h_v = the volumetric heat transfer coefficient;
         G = the mass velocity.
The correlations are plotted in Fig. 8.2 and illustrate the experimental range of G without extrapolation.

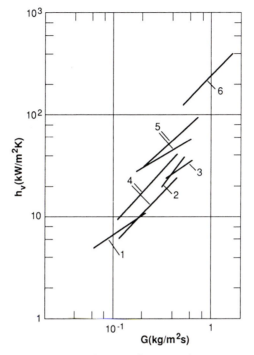

Fig. 8.2. The volumetric heat transfer
coefficient for grains. (1) barley (Boyce,
1966); (2) rice (Wang *et al.*, 1979);
(3) germinated barley (Woods *et al.*, 1992);
(4) barley (Khandker and Woods, 1987);
(5) barley (Woods *et al.*, 1992);
(6) oilseed rape (Woods and Crisp, 1992).

The earliest experiments on heat transfer in grain considered are those
of Boyce (1966). Having brought the grain to moisture equilibrium with the
hot air stream (65–67 °C), the grain, cooled to ambient temperature, was
replaced in the flow. The temperature rise of the grain was then determined
by sampling from the thin layer (25 mm). It is thought likely that the heat
loss during transfer accounts for the low values of h_v observed. It is worth
noting that for all the experiments on grain discussed in this section, on bed
temperatures of around 60 °C and initial grain temperatures of around 20 °C
(ambient) were used.

Wang *et al.* (1979) employed a deep bed (250 mm) technique to determine
h_v for rice. The passage of the temperature wave was observed using
thermocouples within the kernels. The heat transfer coefficient was determined by graphical plots superimposed visually on the solution of Schumann

(1929). The technique was also employed for germinated barley (green malt) in a series of experiments on the drying characteristics of 'malt' for the simulation of malt kilning (Bala, 1983; Bala and Woods, 1984). The correlation is presented in Fig. 8.2 and intersects with the data of Wang *et al.* Khandker and Woods (1987) performed the same experiment on barley. The two correlations plotted result from two methods of fitting the data to the solution of Schumann.

The remaining correlations are for data obtained on thin layers. In their work on the cooling of packed beds of grains and vegetables, Gan (1989) and Gan and Woods (1989) re-examined the data on packed bed heat transfer coefficient. As will become apparent in the next section, the deep bed data processed through the Schumann plots fall well below main stream heat and mass transfer data for packed beds. By considering the manner in which hot air is cooled through the grain bed, it became clear that the relative humidity of the air would rise in its passage through the bed. In deep bed experiments it is therefore difficult to achieve isomoisture conditions.

In view of this, experiments were performed on a thin layer of barley, employing both infra-red pyrometry and thermocouple insert techniques for grain temperature measurement (Gan, 1989; Woods *et al.*, 1992). In the thermocouple insertion technique, the instrumented grains were located 13 mm from the bed entry in a bed of depth 35 mm. The temperature of air adjacent to the instrumented grains was recorded by shielded thermocouples. The pyrometric technique employed the same bed depth with the exit temperature of the grain measured by pyrometer and adjacent air temperature monitored by shielded thermocouples. Elimination of deep bed moisture transfer effects gave higher values of h_v as shown in Fig. 8.2. Furthermore, the pyrometric technique gave higher values of h_v than the thermocouple insert method.

Recent work on the drying properties of oilseed rape (Crisp, 1991) have included the measurement of heat transfer coefficient (Woods and Crisp, 1992). Improvements in technique, whereby the grain temperature at the bed inlet is observed by pyrometer and through thermocouple inserts in grains, has simplified the analysis of data. The simplification arises from the first layer of grains seeing a constant air temperature. There may be a difference in the heat transfer coefficient for the first layer of grain due to the difference in geometry as compared with grains within the bed. This is not quantified and is a penalty incurred by the pyrometric technique. The h_v data for oilseed rape, as shown in Fig. 8.2, are considerably higher than those for barley due to the greater surface area available for the smaller seeds. This effect is better accounted for by the use of dimensionless groups as presented in the next section.

In all these experiments the value of h_v is linearly related to the value of grain specific heat and bulk density. In the work of Bala and Woods (1984) for germinated barley, Khandker and Woods (1987) for barley and Woods and Crisp (1991) for oilseed rape, the values of specific heat were determined for the sample. In the other work on barley (Gan, 1989; Woods *et al.*, 1991) the specific heat values of Boyce (1966) were employed. It is not clear from Wang *et al.* which value of C_p was used. In determining h_v, the gravimetric heat transfer coefficient is first determined and then multiplied by the bulk density. The bulk density measurement in each experiment is therefore very relevant, although there is no problem in correcting for this factor. Volumetric heat transfer coefficient has been used in preference to gravimetric in view of previous conventions.

The history of the measurement of heat transfer coefficient in grain beds is one of increasing estimates as techniques reducing moisture transfer effects have developed. For this reason, the thin layer data giving the higher observed values of h_v are considered for comparison with mainstream heat and mass transfer data in the next section.

General packed bed data

There are many other packed bed situations where data on heat and mass transfer coefficient between particle and fluid have been gathered. This is often for synthetic particles of controlled geometry forming a packed bed reactor. The use of dimensionless groups enables these data to be compared and unified. This approach is employed in the comparison of the grain data with general packed bed results.

It is useful at this stage to consider the range of Reynolds numbers encountered in the deep bed drying and cooling of agricultural crops. From general experience and the recommendations in Anon (1983a, b, c) and Anon (1979), the following typical ranges have been estimated:

$$\text{Cooling vegetables and fruits} \quad Re = 10^2 - 10^4$$
$$\text{Grain drying} \quad Re = 10 - 10^2$$
$$\text{Grain cooling} \quad Re = 1 - 10$$

These are very general indicators as in reality there are no absolute limits on air velocities. However, they are useful in appreciating the relevance of the data to be presented.

Wakao and Kaguei (1982) have brought together much of the data in this area for both heat and mass transfer. Selectively examining the literature and correcting the transfer coefficient for the axial dispersion effect at low Reynolds number, the following correlation was proposed:

$$Nu = 2 + 1.1 \, Re^{0.6} Pr^{1/3} \qquad [8.2]$$

where Nu, Re and Pr are the Nusselt, Reynolds and Prandtl numbers respectively.

This was based on heat transfer data largely for spherical particles in air

with a Reynolds number range of 15 to 8500. It was confirmed by the effective
correlation of mass transfer data for gases and liquids by the analogous
equation

$$Sh = 2 + 1.1 \; Re^{0.6} \; Sc^{1/3} \qquad\qquad [8.3]$$

where Sh = the Sherwood number and Sc = the Schmidt number.

The data were again largely for spheres and the gas phase results covered
the Reynolds number range of 3 to 9900. The liquid phase results cover a
lower range of Reynolds numbers, but the far higher Schmidt numbers give
Sherwood numbers as high as 10^3. It is worth noting that, although for mass
transfer experiments, Sherwood numbers as low as 3 to 4 have been observed,
in heat transfer experiments no observed Nusselt numbers fall below 10. This
is on the basis of values corrected for axial dispersion.

Equation [8.2] is presented in Fig. 8.3, where the solid lines indicate the
range of validity without extrapolation. The work of Gamson *et al.* (1943)
is included, as it has often been referred to in grain drying work. Although
Gamson *et al.* propose correlations at low Reynolds numbers, their results
do not appear to extend below 10^2. Most grain drying applications relate
to Reynolds numbers below 10^2. The correlation of Wilke and Hougen
(1945) is included as it covers part of the Reynolds number range of interest

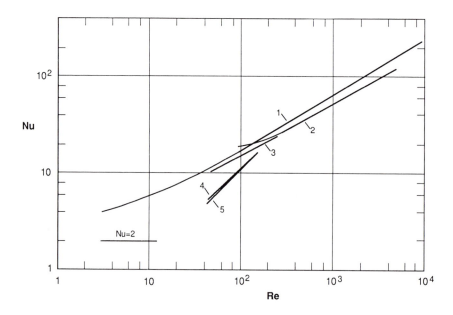

Fig. 8.3. Dimensionless correlations. (1) Wakao and Kaguei, 1982; (2) Gamson *et al.*, 1943; (3) Wilke and Hougen, 1945; (4) barley (Woods *et al.*, 1992) ; (5) oilseed rape (Woods and Crisp, 1992).

in grain drying and is one of the data sets correlated by Wakao and Kaguei.

Inserting the pyrometric measurements of heat transfer coefficient for barley and oilseed rape in dimensionless form in Fig. 8.3, we can see that they still fall below the other correlations. Further consideration is given to this discrepancy in the next section.

In order to convert the values of h_v and G into Nusselt number and Reynolds number, a characteristic dimension and a measurement of surface area were required. The assumption of a spherical geometry for oilseed rape enabled the diameter and surface area to be readily calculated from the bulk density, thousand grain weight and seed specific gravity, as determined using a picnometer. This gave a diameter of 1.93 mm and surface area per unit bulk volume of $1720 \, m^2/m^3$. The diameter was also measured by sieving over a set of square meshes and averaging, giving a value of 1.97 mm (Crisp, 1991).

The diameter and surface area per unit bulk volume of the more geometrically complicated barley grain were calculated as 4.35 mm and $884 \, m^2/m^3$, assuming spheres of equal volume (Gan, 1989). For non-spherical seeds, the problem of specifying a characteristic dimension and surface area is addressed in the section 'Grain diameter and surface area' below.

Axial thermal dispersion

The tortuous nature of the flow path through a grain bed gives rise to considerable velocity variations for each fluid element. The mixing of fluid elements through differing and oscillating relative velocities in the flow or axial direction results in a dispersion of heat. This is in addition to the heat transfer by the mean motion of the fluid. Wakao and Kaguei (1982) put forward the following equation describing axial thermal dispersion:

$$\frac{k_{ax}}{k_a} = \frac{k_o}{k_a} + \delta \, Re \, Pr \qquad [8.4]$$

where k_o and k_a are the thermal conductivities of the quiescent bed and air respectively.

A value for δ of 0.5 is recommended by Wakao and Kaguei; however, the experimental observations of Yagi *et al.* (1960) indicated a value of 0.7 to 0.8.

The effect of axial heat dispersion becomes significant relative to particle to fluid heat transfer coefficient at low flow rates. It is therefore worth considering the effect to explain the low observed values of Nusselt number in grain beds where Reynolds numbers are relatively low. The results for barley (Gan, 1989; Woods *et al.*, 1992) and oilseed rape (Crisp, 1991; Woods and Crisp, 1992) have been analysed in this way.

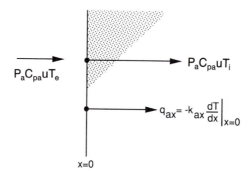

Fig. 8.4. Energy flows at the bed interface.

In Fig. 8.4, the energy flows at the grain bed inlet are illustrated. Equating the energy flow to the bed with the sum of the energy flows by convection and axial dispersion across the plane $x = 0$ gives what is sometimes referred to as the Danckwerts (1958) boundary condition:

$$T_e - T_i = - \frac{k_{ax}}{\rho_a C_{pa} u} \frac{\partial T}{\partial x} \bigg|_{x = 0} \quad\quad [8.5]$$

where T_e and T_i = the air temperatures external to and at the interface of the grain bed;
   ρ_a and C_{pa} = the density and specific heat of air;
   u = the superficial air velocity.

The identical equation applies at the exit to the bed. In physical terms, the air temperature within the bed is lower at the interface due to axial thermal dispersion assisting axial convection. This gives rise to an underestimate of heat transfer coefficient.

Following the analysis of Schumann (1929) and considering the effect of transient air temperature $\partial T / \partial t$ to be negligible gives, at the interface:

$$\frac{\partial T}{\partial x} = - \frac{ha_v}{\rho_a C_{pa} u} (T_i - \Theta) \quad\quad [8.6]$$

$$\frac{\partial \Theta}{\partial t} = - \frac{ha_v}{\rho_b C_{pg}} (T_i - \Theta) \quad\quad [8.7]$$

where T and Θ = air and grain temperatures;
   h = the surface heat transfer coefficient;
   a = the bed surface area per unit volume;
   ρ_b = the grain bulk density; and
   C_{pg} = the grain specific heat.

From equations [8.5], [8.6] and [8.7] we can write:

$$\frac{\partial \Theta}{\partial t} = - \frac{a_v k_a}{d \rho_b C_{pg}} \frac{Nu}{(1 + \phi \, Nu)} (T_e - \Theta) \qquad [8.8]$$

Not including the axial dispersion effect produces the equation:

$$\frac{\partial \Theta}{\partial t} = - \frac{a_v k_a}{d \rho_b C_{pg}} Nu' \, (T_e - \Theta) \qquad [8.9]$$

where Nu' is the apparent Nusselt number.
Comparing these equations we can write:

$$Nu' = \frac{Nu}{1 + \phi \, Nu} \qquad [8.10]$$

Rearranging gives the true Nusselt number, Nu as:

$$Nu = \frac{Nu'}{1 - \phi \, Nu'} \qquad [8.11]$$

where:

$$\phi = \frac{(a_v d)(k_{ax}/k_a)}{Re^2 \, Pr^2} \qquad [8.12]$$

This correction technique applies to the results for oilseed rape, where grain temperature was observed pyrometrically at the bed inlet and for barley, where observations were made at the bed exit.

From the work of Timbers (1975) and Moysey *et al.* (1977) the thermal conductivity of an unventilated bed of oilseed rape at a moisture content of 4.2% (Woods and Crisp, 1992) is typically 0.11 W/mK. For air at the bed entry temperature of 60 °C, from Eqn [8.4]:

$$\frac{k_{ax}}{k_a} = 3.82 + \delta \, Re \, Pr \qquad [8.13]$$

There are little data on the thermal conductivity of quiescent beds of barley. Mohsenin (1980) presents a single value of 0.15 at 12.2% moisture content. From the work of Oxley (1944), Babbit (1945) and Moote (1953), the thermal conductivity of an unventilated bed of wheat at a moisture content of 4.9% (Woods *et al.*, 1992) is typically 0.15 W/mK. This value is taken as the best estimate available for barley. For air at the mean of the typical bed entry (60 °C) and typical ambient (20 °C):

$$\frac{k_{ax}}{k_a} = 5.49 + \delta \, Re \, Pr \qquad [8.14]$$

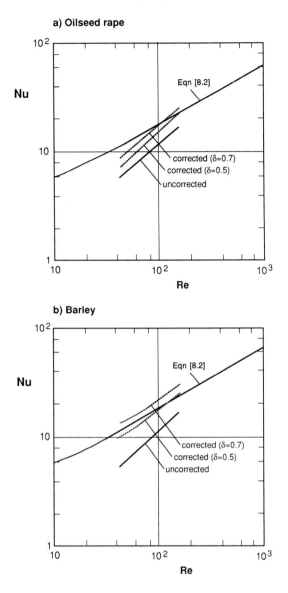

Fig. 8.5. Dimensionless data correlations corrected for axial thermal dispersion and compared with Equation [8.2] from Wakao and Kaguei (1982). Barley data: Woods *et al.*, 1992; oilseed rape data: Woods and Crisp, 1992.

The effect of the correction for thermal dispersion is shown in Fig. 8.5. In general, it brings all the data closer to the correlation of Wakao and Kuguei (1982). For barley, the correction also improves the slope of the Nu versus Re plot. The larger value of the quiescent bed conductivity, k_o produces this effect. The lower value of quiescent bed conductivity for oilseed rape is observed by both Timbers (1975) and Moysey *et al.* (1977) and the slope of the Nu versus Re plots is not reduced (Fig. 8.5(a)).

The axial thermal dispersion correction has brought the observed results for grains into good agreement with general packed bed data.

Grain diameter and surface area

In order to compare the heat transfer coefficient in a grain bed with data for other packed bed materials, it is necessary to specify a characteristic dimension for use in dimensionless parameters. It is also necessary to convert the heat transfer on a gravimetric basis, as measured in most experimental techniques, to the heat transfer coefficient based on particle surface area. This requires a specific surface area measurement related to the material weight.

Before tackling the questions of characteristic dimension and surface area, it is useful to consider some of the basic measurements of size available.

Available measurements

The most direct measurement of grain size is the combination of length a, width b, and depth c of a seed. These are usually defined in descending order of magnitude. A mean diameter is defined by:

$$d = \sqrt[3]{abc} \qquad [8.15]$$

If the grain can be considered as an ellipsoid, then d is the diameter of a sphere of equivalent volume. This is a useful technique but laborious for large samples.

The simplest and perhaps most useful measure of grain size is the average weight of a grain w_g. This is obtained relatively quickly by counting and weighing a reasonably large number of grains. The term 'thousand grain weight', W_{1000}, is well accepted but does not necessarily indicate a sample size of one thousand.

In determining grain size from grain weight the density ρ_g of the grain material itself is required. This can be measured by means of a picnometer or fluid displacement technique (Mohsenin, 1970). The grain density ρ_g, bulk density ρ_b, and void ratio ϵ are related by the equation:

$$\epsilon = 1 - \rho_b/\rho_g \qquad [8.16]$$

The bulk density of a sample under test is readily available. Equation [8.16] then gives the void ratio of the bed.

The subsequent analysis required w_g, ρ_b and ρ_g (or ϵ) to be specified. In situations where the more involved grain density measurement is not available, an assumption for ρ_g or ϵ is required. The assumption of a constant density value is more appropriate as this will not vary with the packed bed filling procedure which affects bulk density and void ratio.

Grain surface area

Assuming the geometric proportions and specific gravity for all grains of a given crop are constant, the surface area of a grain can be written:

$$a_g = s(w_g)^{2/3} \qquad [8.17]$$

where s = a constant with units of $m^2/kg^{2/3}$;
        a_g and w_g = the surface area and weight of a single grain.

Although the fullness of the grain and the relative proportions of component materials may vary, Eqn [8.17] is considered to be a useful working hypothesis. The weight of a grain, w_g, is readily obtained from the thousand grain weight, W_{1000}. The surface area of grain per unit weight is then:

$$\frac{a_g}{w_g} = \frac{s}{w_g^{1/3}} \qquad [8.18]$$

The heat transfer coefficient per unit surface area is given by:

$$h = \frac{h_g \, w_g^{1/3}}{s} \qquad [8.19]$$

where h_g = the gravimetric heat transfer coefficient.

Table 8.1. The specific surface area of grains.

	W_{1000} (g)	ρ_b (kg/m^3)	s (m^2/kg$^{2/3}$)
Wheat[a] (spring)	34.6	768.9	0.0486
Wheat[a] (winter)	39.6	768.9	0.0539
Wheat[a] (average)	37.2	768.9	0.0513
Barley[a]	34.3	768.9	0.0625
Oats[a]	41.0	410.1	0.0922
Corn[a]	336.5	820.1	0.0665
Oilseed rape[b]	4.93	726.1	0.0404[c]

[a] from Bakkar-Arkema et al., 1971.
[b] from Crisp, 1991.
[c] calculated assuming a sphere, $s = (36\pi/\rho_g^2)^{1/3}$, taking ρ_g from Table 8.2.

The values of s for a number of crops can be obtained from the experimental data of Bakker-Arkema *et al.* (1971). By coating grains with a fine powder of nickel particles and weighing, the surface area was determined. The results are recalculated to give the specific surface area, s and presented in Table 8.1. The value for barley is used in the calculation of surface heat transfer coefficient later in this section.

Diameter of an equivalent sphere

In the correlating equation (Eqn [8.2]) of Wakao and Kuguei (1982) for existing chemical engineering data, 70–80% of the data were obtained using beds of spheres. In order to use this correlation for grain, the appropriate dimension, equivalent to the diameter of a sphere, must be identified. There are a number of ways of defining equivalent diameter.

Equivalent surface area

This is the diameter of a sphere with the same surface area as a grain and can be written:

$$d_a = \left(\frac{s}{\pi}\right)^{1/2} (w_g)^{1/3} \qquad [8.20]$$

Gamson *et al.* (1943) and Wilke and Hougen (1945) both use this diameter as their characteristic dimension.

Equivalent volume

The diameter of the sphere of the same volume can be written:

$$d_v = \left(\frac{6}{\pi \rho_g}\right)^{1/3} (w_g)^{1/3} \qquad [8.21]$$

Equivalent surface to volume ratio

The diameter of the sphere with equivalent surface to volume ratio can be written:

$$d_r = \frac{6}{\rho_g s} (w_g)^{1/3} \qquad [8.22]$$

This is clearly relevant to the internal diffusion of moisture and heat, as it is a measure of the mean diffusion distance to the surface. It is suggested that d_r is appropriate to transient internal diffusion analysis where spherical geometry is assumed.

Table 8.2. Physical properties of grains.

	Moisture content (% wet basis)	W_{1000} (g)	ρ_g (kg/m³)	ρ_b (kg/m³)	ϵ	$(abc)^{1/3}$ (mm)
Wheat[a] [7]	7.72 (9.74)	43.4 (13.5)	1420 (0.575)	801 (1.31)	0.436 (1.73)	4.02 (5.44)
Barley[a] [6]	7.77 (3.61)	45.0 (14.9)	1390 (1.17)	605 (5.21)	0.566 (3.14)	4.64 (5.62)
Oats[a] [5]	8.66 (1.55)	34.0 (9.68)	1370 (0.953)	464 (13.2)	0.661 (6.62)	4.39 (7.59)
Corn[a] [1]	6.70	346.4	1290	745	0.423	16.2
Oilseed rape[b] [1]	4.2	4.93	1310	726	0.446	1.97[c]

[ ]   number of varieties.
( )   standard deviation between varieties as a percentage of the average value.
[a]   from the data of Goss, presented in Mohsenin, 1970.
[b]   from Crisp, 1991.
[c]   determined from sieved size distribution.

Table 8.3. Calculated diameters of grains (mm).

	d_a	d_v	d_r	D_h	$(abc)^{1/3}$
Wheat	4.49	3.88	2.89	1.49	4.02
Barley	5.02	3.95	2.46	2.13	4.64
Oats	5.55	3.62	1.54	2.00	4.39
Corn	10.2	8.00	4.91	2.46	16.2
Oilseed rape	1.93[a]	1.93[a]	1.93[a]	1.04	1.97[b]

[a] the diameters are identical as the specific surface area s was calculated assuming a sphere (Table 8.1).
[b] determined from sieved size distribution. (From Crisp, 1991.)

The data of Goss (unpublished) from Mohsenin (1970), are particularly comprehensive and provide values for w_g, ρ_b and ρ_g, as shown in Table 8.2. It should be noted that the moisture content for these data is generally low. Based on these values, and the surface area data in Table 8.1, the three equivalent diameters are calculated and presented in Table 8.3. For a given crop the diameter varies considerably for the different assumptions. The variation is greatest for the least spherical seeds, which generally have the highest specific surface area and void ratio. Corn, on the other hand, has a low void ratio with quite a high specific surface area. Pabis and Henderson (1961) described corn geometry assuming a brick shape, which offers the possibility of such a compact packing of the bed.

In view of the variation between the three equivalent diameters for the less spherical crops, some physical basis for the selection and use of an appropriate diameter is required.

Diameter and velocity in the void space

The physical phenomena that control heat transfer coefficient are those taking place in the air flow in the spaces between the grains. Blake (1922) recognized the importance of interstitial velocity and a dimension characterizing the void space. His work was followed by that of Carman (1937), Ergun (1952) and Carman (1956), who further utilized these principles to describe the flow resistance of packed beds. Applying these principles to heat transfer data, a characteristic dimension of the flow is therefore better determined from the geometry of the air spaces. The diameter of a cylindrical tube with the same ratio of surface area to volume as the cavity formed by the grains is given by:

$$D_h = 4 \frac{V}{A} \qquad [8.23]$$

where   V = the volume of the cavity;
        A = the surface area of the grains.

Writing this in terms of our measurable parameters we have:

$$D_h = \frac{4\epsilon \; w_g^{1/3}}{\rho_g(1 - \epsilon) \; s} \qquad [8.24]$$

This leads to a most useful result. Combining Eqns [8.22] and [8.24] gives:

$$D_h = \frac{2\epsilon}{3(1 - \epsilon)} \; d_r \qquad [8.25]$$

The characteristic dimension of the flow D_h is proportional to the equivalent diameter d_r. It also takes into account the effect of void ratio.

    Similarly the characteristic velocity can be taken as the mean velocity in the air cavity between the grains. This is related to the superficial air velocity, u by:

$$v = u/\epsilon \qquad [8.26]$$

For a sphere, d_r is by definition equal to d and the void ratio ϵ should lie in the region of 0.395 (orthorhombic lattice) to 0.476 (cubic lattice). Taking ϵ for a bed of spheres as 0.44, then from Eqn [8.25]:

$$d = 1.91 \; D_h \qquad [8.27]$$

and from Eqn [8.26]:

$$u = 0.44 \; v \qquad [8.28]$$

The recommended correlating equation of Wakao and Kaguei (1982) for heat transfer coefficient (Eqn [8.2]), which is based largely on data for spheres, can therefore be written:

$$\frac{hD_h}{k_a} = 1.048 + 0.519 \left(\frac{vD_h}{\nu_a}\right)^{0.6} Pr^{1/3} \qquad [8.29]$$

This equation has now been written in terms of the characteristic velocity and dimension for the air cavity. It is useful to rewrite this in terms of the equivalent grain diameter d_r and the superficial velocity u using Eqns [8.25] and [8.26]:

$$Nu_r = \frac{(1 - \epsilon)}{\epsilon} \left(1.57 + \frac{0.611}{(1 - \epsilon)^{0.6}} \; Re_r^{0.6} \; Pr^{1/3}\right) \qquad [8.30]$$

For a void ratio ϵ of 0.44 the equation is identical with the original correlation for spheres, Eqn [8.2]. It should be noted that Nu_r and Re_r are based on the equivalent diameter d_r defined in Eqn [8.22].

Further analysis of the barley data

The description of the data for the near spherical oilseed rape does not require the added complexity of Eqn [8.30]. However, from the data in Tables 8.1 and 8.2, barley has a void ratio and surface area per unit weight value far higher than a spherical particle. Since detailed data on void ratio and surface area are not available for the barley sample used for the heat transfer experiment, the values presented by Mohsenin (1970) in Table 8.2 are used together with the surface area measurement of Bakker-Arkema *et al.* (1971) in Table 8.1. These give a void ratio ϵ of 0.566, a d_r of 2.46 mm and surface area per unit volume a of $1063 \, m^2/m^3$.

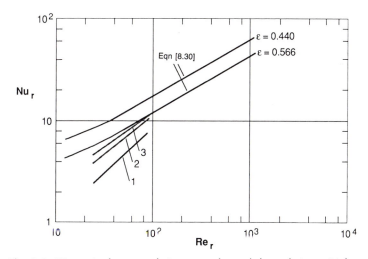

Fig. 8.6. Dimensionless correlations transformed through interstitial velocity and hydraulic duct diameter, and compared with the data for barley (Woods *et al.*, 1992). (1) data uncorrected for axial thermal dispersion; (2) data corrected ($\delta = 0.5$); (3) data corrected ($\delta = 0.7$).

The Nusselt and Reynolds numbers are recalculated using d_r and a. These are then compared with Eqn [8.30] for $\epsilon = 0.566$ in Fig. 8.6. The axial diffusion correction is also applied using a modified form of Eqn [8.13]. Assuming that Eqn [8.13] is valid for a bed of spheres with a void ratio of 0.44, and writing the Reynolds number in terms of diameter d_r, we have:

$$\frac{k_{ax}}{k_a} = \frac{k_o}{k_a} + \frac{0.56}{1 - \epsilon} \delta \, Re_r \, Pr \qquad [8.31]$$

Values of Nu, corrected for axial thermal dispersion taking $\delta = 0.5$ and 0.7 as before, are also presented in Fig. 8.6.

These results show a similar pattern to the previous values in Fig. 8.5. The correction for axial dispersion brings the data correlation close to the transformed correlating equation of Wakao and Kaguei (1982). All values are lower due to the effects of d_r and a on Nusselt and Reynolds numbers.

The transformed version of Wakeo and Kaguei's correlating equation gives as good an agreement as the original form, whilst incorporating the effects of ϵ, d_r and a. In physical terms, the transformed equation incorporates the interstitial velocity effect through void ratio and the surface area to grain volume effect through d_r and a. In this way the analysis offers a theoretical basis for the selection of an equivalent diameter for non-spherical grains.

Internal conduction

In their investigation of the air–grain temperature relationship, Pabis and Henderson (1962) measured the temperatures at the centre and the surface of a maize kernel during drying. They observed that after three to four minutes no difference could be detected between these temperatures. The sensitivity of their temperature measurement was 0.6 °C. The difference in temperature between the air and the grain was of the order of 1.2 °C after 4 min. This clearly indicated internal resistance to heat transfer to be relatively small.

Recent work concentrating on the measurement of heat transfer coefficient has included the effect of internal conduction (Sokhansanj and Bruce, 1987; Gan, 1989). It is therefore useful to evaluate the relative magnitude of internal and external heat transfer resistance in the light of more recent work.

Thermal conductivity of a grain

The data on grain thermal conductivity have been almost totally concerned with the conduction of grain as a bulk material and not within the grain material itself. Perhaps this has not been made sufficiently clear in some data compilations (Anon, 1981; Anon, 1989). The thermal conductivity of wheat in bulk is plotted against moisture content in Fig. 8.7, after the data of Oxley (1944), Babbitt (1945) and Moote (1953). Values for the thermal conductivity of the grain material itself are also plotted from the data of Chubik and Maslow (1970), as presented by Sitkei (1986). At 15–20% moisture content the thermal conductivity of the grain material itself is of the order three times that of the grain bed as a whole. The values used by Sokhansanj and Bruce (1987) and Gan (1989) for internal conduction were of the same order as the bulk material. The heat transfer correlation used by Sokhansanj and Bruce gave lower values than Gamson et al. (1943), which in turn are lower than Wakao and Kuguei (1982). The introduction of internal heat conduction

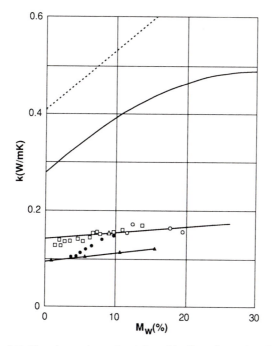

Fig. 8.7. The thermal conductivity of bulk grain and grain material. Bulk wheat, o – Oxley (1944); ∆ – Babbitt (1945); □ – Moote (1953). Bulk oilseed rape, ● – Timbers (1975); ▲ – Moysey *et al.* (1977). For wheat grain material, ——— Chubik and Maslow (1970); - - - - calculated from Wakao and Kato (1969).

analysis to explain the effect of the low correlated values of heat transfer coefficient may therefore not have been necessary. In the work of Gan, the internal conduction analysis was introduced in an attempt to improve the Schumann plot technique. Due to the moisture transfer problems in deep beds already discussed, thin layer methods have superseded this work.

Based on the analysis of Wakao and Kato (1969), it is possible to estimate the thermal conductivity of grain material from the conductivity of the bulk material. The method is based on a theoretical solution for conduction and radiation in an orthorhombic lattice of spheres, having a void ratio of 0.395. The radiation effect is secondary but was included taking diameter, emissivity and typical temperature as 4.35 mm, 0.9 and 40 °C respectively. The technique is applied to the data for bulk wheat in Fig. 8.7, giving the estimated grain material thermal conductivity shown. The predicted value is approximately 30% higher than the data of Chubik and Maslow (1970). Considering the assumption of orthorhombic spheres and stationary air, the level of agreement gives useful support to the data.

Data for the thermal conductivity of wheat flour (Reidy, 1968) in the range of 0.45 to 0.69 W/mK at 8.8% moisture content tend to confirm these values. Hallstrom *et al.* (1988) quote thermal conductivities for carbohydrate, protein and oil of 0.58, 0.20 and 0.18 W/mK respectively. This also supports the observations for wheat grains and suggests that seeds of a higher oil content would have lower thermal conductivities. Data for the bulk thermal conductivity of oilseed rape from Timbers (1975) and Moysey *et al.* (1977) shown in Fig. 8.7 are clearly lower than the values for wheat.

From these observations it can be concluded that the internal thermal conductivity of wheat at moisture contents in the range 10–20% is of the order 0.4 to 0.5 W/mK.

Relative internal and external resistance

Having examined the value of thermal conductivity for grain material, it is now possible to examine the conditions under which internal conduction limits heat transfer.

The heat flux across the boundary layer can be written:

$$q = h \, \Delta T_a \qquad [8.32]$$

where the heat transfer coefficient, h has already been shown to be described by:

$$Nu = \frac{hd}{k_a} = 2 + 1.1 \, Re^{0.6} \, Pr^{1/3} \qquad [8.33]$$

In order to describe the internal conduction in a similar way, a typical conduction distance y is considered. Then:

$$q = \frac{k_g}{y} \, \Delta T_g \qquad [8.34]$$

where k_g is the thermal conductivity of the grain material. A representative value of y is the average distance of the elements of a sphere from the surface, giving a value of d/8. From Eqns [8.32], [8.33] and [8.34], the relative resistance of the internal conduction to the external convection is given by:

$$\frac{\Delta T_g}{\Delta T_a} = \frac{1}{8} \frac{k_a}{k_g} (2 + 1.1 \, Re^{0.6} \, Pr^{1/3}) \qquad [8.35]$$

It is interesting to evaluate this ratio under high Reynolds number conditions. From inspection of the commercial literature for grain drying a superficial velocity of 0.3 m/s is at the top end of the working range. Based on the work already described, the following values are taken:

at 40 °C: $k_a = 0.027 \, \text{W/mK}, \quad \nu_a = 1.7 \times 10^{-5} \text{m}^2/\text{s}, \quad \text{Pr} = 0.71,$
$\qquad k_g = 0.45 \, \text{W/mK}, \quad d = 0.004.$

giving: $\quad \text{Re} = 70.5, \quad \text{Nu} = 14.6, \quad \text{and } \Delta T_g / \Delta T_a = 0.110.$

This clearly demonstrates that even under these extreme conditions the internal grain temperature difference is a second order effect at 11%. For larger grains like corn, where the Reynolds number is approximately doubled for the same velocity, the ratio only rises to 16%. During the early stages of heat transfer, when the surface of the grain is being heated, the typical conduction distance will be even less than $d/8$, further reducing internal resistance. This work indicates the assumption of uniform grain temperature to be valid for most grain drying situations.

Conclusion

The data on convective heat and mass transfer in packed beds have been reviewed and, based on this, equations describing the particle-to-air convective heat transfer coefficient are proposed. The initial objective of refining the method by which we determine crop temperature during drying and cooling has been achieved. The main points in the conclusion are summarized as follows.

1.   The correlating equations of Wakao and Kaguei (1982) for the convective heat and mass transfer coefficients (Eqns [8.2] and [8.3]) form the best available quantitative summary of the data available. This work is based largely on data for spheres.
2.   Correcting for the axial thermal dispersion at the interface, as described by the Danckwerts (1958) boundary condition, brings the data for barley and oilseeds into good agreement with the correlation of Wakao and Kaguei.
3.   The non-spherical nature of grains has been allowed for by modifying the correlation of Wakao and Kaguei. This was achieved by considering the hydraulic radius of the void space and the interstitial velocity to be the characteristic velocity and dimension respectively. Since the hydraulic radius is simply related to the equivalent spherical diameter based on surface-to-volume ratio, a convenient form of the correlation results and is presented in Eqn [8.30]. A theoretical basis for the selection cf an equivalent diameter for non-spherical grains has been demonstrated.
4.   The surface area of grains is taken from the data of Bakker-Arkema *et al.* (1971), who used a particle coating technique for its measurement. A constant is derived for each type of grain relating the surface area to the single grain weight to the power ⅔.
5.   In calculating grain diameter and void ratio, the recommended approach

in the absence of a picnometer, is to measure bulk density and thousand grain weight and assume a density for the grain material itself, based on previous work. This is preferable to assuming void ratio.

6.   Internal conduction resistance for grains is shown to be a second-order effect in most drying and cooling situations. This effect has been considered for high-moisture crops by Woods (1990), who concluded it to be significant for larger vegetables and fruits. The ambiguity between bulk grain and grain material thermal conductivity has been clarified.

These conclusions, in combination, provide a basis for the description of convective processes in packed bed drying and cooling of crops.

Nomenclature

a_v	bed surface area per unit volume
A	surface area of the packed bed
a_g	single grain surface area
a, b, c	grain length, width and depth
C_p	specific heat at constant pressure
d	characteristic grain diameter
d_a	value of d based on the equivalent area to a sphere
d_v	value of d based on the equivalent volume to a sphere
d_r	value of d based on the equivalent area/volume ratio to a sphere
D_h	hydraulic diameter of the void space
e, f	correlating coefficients
G	mass velocity
h	surface heat transfer coefficient
h_g	gravimetric heat transfer coefficient
h_v	volumetric heat transfer coefficient
k	thermal conductivity
k_{ax}	effective thermal conductivity of a bed
k_o	thermal conductivity of a quiescent bed
Nu	Nusselt number, hd/k
Nu_r	Nusselt number based on d_r and measured grain surface area
Nu'	apparent experimental Nusselt number
Pr	Prandtl number
q	heat flux
Re	Reynolds number, ud/ν
Re_r	Reynolds number based on d_r
s	specific surface area of a single grain ($a_g = s\ w_g^{2/3}$)
Sc	Schmidt number
Sh	Sherwood number
t	time
T	air tempereature
T_e	air temperature external to the bed

T_i	air temperature at the inlet and outlet interfaces to the bed
ΔT_a	temperature difference across the boundary layer
ΔT_g	temperature difference across the grain
u	superficial air velocity
v	interstitial air velocity
V	void volume of the packed bed
w_g	single grain weight
W_{1000}	thousand grain weight
x	spatial coordinate in the flow direction
y	typical heat conduction distance within a grain
δ	coefficient for axial thermal dispersion
ϵ	void ratio
θ	grain temperature
ρ	density
ϕ	dimensionless group for axial thermal dispersion correction
ν	kinematic viscosity

Subscripts

a	air
g	grain material
b	grain bulk material

References

Anderson, R. J. (1972) Commercial concurrent-flow heating – counterflow cooling grain drier – Anderson Model. American Society of Agricultural Engineers, Paper Number 72-846

Anon (1979) Refrigerated storage of fruits and vegetables. Ministry of Agriculture, Fisheries and Food, Reference Book 324.

Anon (1981) Thermal properties of foods. In: American Society of Heating, Refrigerating and Air-conditioning Engineers, *Fundamentals Handbook*, ch. 31.

Anon (1983a) High temperature grain drying. Ministry of Agriculture, Fisheries and Food, Booklet 2417.

Anon (1983b) Bulk grain dryers. Ministry of Agriculture, Fisheries and Food, Booklet 2416.

Anon (1983c) Grain storage methods. Ministry of Agriculture, Fisheries and Food, Booklet 2415.

Anon (1989) *American Society of Agricultural Engineers Standards*. St Joseph, MI, USA.

Babbit, J. D. (1945) The thermal properties of wheat in bulk. *Canadian Journal of Research F* 23, 388–401.

Bakker-Arkema, F. W., Rosenau, J. R. and Clifford, W. H. (1971) The effect of grain surface area on the heat and mass transfer rates in fixed and moving beds of biological products. *Transactions of the American Society of Agricultural Engineers* 14, 864–7.

Bala, B. K. (1983) 'Deep bed drying of malt.' Ph.D. Thesis, University of Newcastle upon Tyne.

Bala, B. K. and Woods, J. L. (1984) Simulation of deep bed malt drying. *Journal of Agricultural Engineering Research* 30(3), 235–44.

Baughman, G. R. (1971) 'Experimental study and simulation of a concurrent flow grain dryer.' Ph.D. Thesis, Ohio State University.

Blake, F. C. (1922) The resistance of packing to fluid flow. *Transactions of the American Institute of Chemical Engineers* 14, 415–21.

Boyce, D. S. (1966) 'Heat and moisture transfer in ventilated grain.' Ph.D. Thesis, University of Newcastle upon Tyne.

Carman, P. C. (1937) Fluid flow through granular beds. *Transactions of the Institution of Chemical Engineers* 15, 150–66.

Carman, P. C. (1956) *Flow of Gases through Porous Media*. Butterworths, London.

Chubik, L. and Maslow, A. (1970) *Spravochnik po Teplofizicheskim Kharakteristkam Pishchevikh Productov (Manual of the Thermophysical Characteristics of Food Products)*. Izd. Pishchevaya Prom., Moscow.

Crisp, J. (1991) 'Drying properties of rapeseed.' Ph.D. Thesis, University of Newcastle upon Tyne (in preparation).

Danckwerts, P. V. (1958) Continuous flow systems: Distribution of residence times. *Chemical Engineering Science* 2(1), 1–13.

Ergun, S. (1952) Fluid flow through packed columns. *Chemical Engineering Progress* 48(2), 89–94.

Gamson, B. W., Thodos, G. and Hougen, O. A. (1943) Heat, mass and momentum transfer in the flow of gases through granular solids. *Transactions of the American Institute of Chemical Engineers* 39, 1–35.

Gan, G, (1989) 'Heat and moisture transfer in the deep bed cooling of grain and vegetables.' Ph.D. Thesis, University of Newcastle upon Tyne.

Gan, G. and Woods, J. L. (1989) A deep bed simulation of vegetable cooling. *Proceedings of the Eleventh International Congress on Agricultural Engineering (CIGR)*, Dublin, Ireland. Balkema, V.4, 2301–8.

Hall, G. E. and Anderson, R. J. (1980) Batch internal recycling drier. American Society of Agricultural Engineers, Paper Number 80-3515.

Hallstrom, B., Skjolderbrand, C. and Tragardh, C. (1988) *Heat Transfer and Food Products*. Elsevier, London.

Isaacs, G. W. and Muhlbauer, W. (1975) Possibilities and limits of energy saving in maize grain drying. *Landtechnik* 30(9), 397–401.

Khandker, M. H. K. and Woods, J. L. (1987) The heat transfer coefficient in grain beds. *Agricultural Mechanization in Asia, Africa and Latin America* 18(4), 39–44.

Mohsenin, N. N. (1970) *Physical Properties of Plant and Animal Materials*. Gordon and Breach, New York.

Mohsenin, N. N. (1980) *Thermal Properties of Foods and Agricultural Materials*. Gordon and Breach, New York.

Moote, I. (1953) The effect of moisture on the thermal properties of wheat. *Canadian Journal of Technology* 31, 57–69.

Moysey, E. B., Shaw, J. T. and Lampman, W. P. (1977) Thermal properties of rapeseed. *Transactions of the American Society of Agricultural Engineers* 20, 768–71.

Muhlbauer, W., Scheuermann, A., Maurer, K. and Blumel, K. (1971) Drying of grain maize by the concurrent flow method at high temperatures. *Grundlagen der Landtechnik* 21(1), 1–5.

Nellist, M. E. (1982) Developments in continuous-flow grain driers. *The Agricultural Engineer* Autumn 1982, 74–80.

Oxley, T. A. (1944) The properties of grain in bulk. III. The thermal conductivity of wheat, maize and oats. *Society of Chemical Industry Journal* 63, 53–5.

Pabis, S. and Henderson, S. M. (1961) Grain Drying Theory. II. A critical analysis of the drying curve for shelled maize. *Journal of Agricultural Engineering Research* 6(4), 272–7.

Pabis, S. and Henderson, S. M. (1962) Grain Drying Theory. III. The air/grain temperature relationship. *Journal of Agricultural Engineering Research* 7(1), 21–6.

Reidy, G. A. (1968) 'Thermal properties of foods and methods of their determination.' M.Sc. Thesis, Michigan State University, East Lansing.

Schumann, T. E. W. (1929) Heat transfer: A liquid flowing through a porous prism. *Journal of the Frankline Institute* 208, 405–16.

Sitkei, G. (1986) *Mechanics of Agricultural Materials*, Developments in Agricultural Engineering 8. Elsevier, Amsterdam.

Sokhansanj, S. and Bruce, D. M. (1987) A conduction model to predict grain temperatures in grain drying simulation. *Transactions of the American Society of Agricultural Engineers* 30(4), 1181–4.

Timbers, G. E. (1975) Properties of rapeseed: 1. Thermal conductivity and specific heat . *Canadian Agricultural Engineering* 17(2), 81–4.

Wakao, N. and Kaguei, S. (1982) *Heat and Mass Transfer in Packed Beds*. Gordon and Breach, London.

Wakao, N. and Kato, K. (1969) Effective thermal conductivity of packed beds. *Journal of Chemical Engineering of Japan* 2(1), 24–33.

Wang, C. Y., Rumsey, T. R. and Singh, R. P. (1979) Convective heat transfer coefficient in a packed bed of rice American Society of Agricultural Engineers, Paper Number 79-3040.

Westelaken, C. M. (1977) Concurrent-flow commercial grain driers – The Westelaken Models. American Society of Agricultural Engineers, Paper Number 77-3016.

Wilke, C. R. and Hougen, O. A. (1945) Mass Transfer in the flow of gases through granular solids extended to low modified Reynolds numbers. *Transactions of the American Institute of Chemical Engineers* 41, 445–51.

Woods, J. L. (1990) Moisture loss from fruits and vegetables. *Postharvest News and Information* 1(3), 195–9.

Woods, J. L. and Crisp, J. (1992) The convective heat transfer coefficient in a bed of rapeseed. *Journal of Agricultural Engineering Research*. (In press.)

Woods, J. L., Bala, B. K. and Gan, G. (1992) The heat transfer coefficient in a grain bed. *Drying Technology*. (In press.)

Yagi, S., Kunii, D. and Wakao, N. (1960) Studies on axial effective thermal conductivities in packed beds. *American Institute of Chemical Engineering* 6(4), 543–6.

Chapter 9

Damage Mechanisms in the Handling of Fruits

Margarita Ruiz Altisent

Introduction

The recognition of an increasing and worldwide demand for high quality in fruits and vegetables has grown in recent years. Evidence of severe problems of mechanical damage is increasing, and this is affecting the trade of fruits in European and other countries. The potential market for fresh high-quality vegetables and fruits remains restricted by the lack of quality of the majority of products that reach consumers; this is the case for local as well as import/export markets, so a reduction in the consumption of fresh fruits in favour of other fixed-quality products (dairy in particular) may become widespread. In a recent survey (King, 1988, cited in Dellon, 1989), it appears that, for the moment, one third of the surveyed consumers are still continuing to increase their fresh produce consumption. The factors that appear as being most important in influencing the shopping behaviour of these consumers are taste/flavour, freshness/ripeness, appealing look, and cleanliness.

Research on mechanical damage in fruit and vegetables has been underway for several years. The first research made on physical properties of fruits was in fact directed towards analysing the response to slow or rapid loading of selected fruits (Fridley *et al.*, 1968; Horsefield *et al.*, 1972). From that time on, research has expanded greatly, and different aspects of the problem have been approached. These include applicable mechanical models for the contact problem, the response of biological tissues to loading, devices for detecting damage causes in machines and equipment, and procedures for sensing bruises in grading and sorting.

This chapter will be devoted to the study of actual research results relative to the cause and mechanisms of mechanical damage in fruits (secondarily in vegetables), the development of bruises in these commodities, the models that have been used up to now, and the different factors which have been recognized as influencing the appearance and development of mechanical damage in fruits. The study will be focused mainly on contact-damage – that is, slow or rapid loads applied to the surface of the products and causing bruises. (A bruise is defined as an altered volume of fruit tissues below the skin that is discoloured and softened.) Other types of mechanical damage, like abrasion and scuffing, punctures and cuts, will be also mentioned briefly.

The incidence of mechanical damage

Harvesting, post-harvest handling, packing, transportation and distribution of fruits and vegetables involve numerous mechanical operations and much impact-related flesh bruising. Impact has been recognized as the most important cause of damage (bruising) in fruits. Excessive compression also causes bruising, as do repeated impacts.

The apple is one of the most problematic fruits in relation to mechanical damage. It has also been extensively studied, and some data have been gathered on the percentage of fruits that are bruised during harvesting and grading. It can be as high as 81% of bruised apples during harvesting, 93% after transporting, and 91–95% caused by bagging (Timm *et al.*, 1989), all using manual harvesting systems.

In a study recently made in the Danish market (Kampp, 1990), it was established that only a few of the examined fruit samples met the EC quality standards for the products studied: 18 varieties of apples, and different numbers of varieties of strawberries, carrots, peaches and nectarines. In the retail samples, more than 20% of the strawberries had pressure damage; 20% of the examined peaches and nectarines had pressure or impact damages; and about 95% of the apple samples did not comply with the EC standards for bruises, 55% of the apples having 1–6 bruises per fruit. In addition, it was observed that some of the produce was being sold unripe, having been harvested at too early a stage.

In Spanish production of fruits and vegetables, quality control is being applied by a leading group of commercial companies (Valenciano García, 1990). Apple and pear samples were examined at retail stores; bruise damage was responsible for 50% of the total damage observed (which amounted to 23–35% including diseases, peel, shape, size, peduncle, etc.). In pears, 10–25% of the observed total class-rejection damage was due to bruises. Other products studied included strawberries, lettuce and green peppers. In the case of strawberries, nearly all damage was caused in the field, and was related to overmaturity of the fruits. Iceberg lettuce showed 5.5% bruise damage (from a total damage of 6.5%). In the case of green peppers, 8.5% of the product showed mechanical damage, most of it being caused during field harvesting and transporting processes within Spain (Ruiz Altisent,

1990, unpublished data). Data obtained in some onion grading plants showed that a high proportion of the product is being rejected from the highest-quality (export) class, amounting to 25–35%; similar figures are observed for potatoes. All these results mean there is a real need to improve the systems and the products to avoid very high economic losses from quality reduction and actual product losses. Expert systems are being applied to solve these problems.

Consumer safety is one of the main concerns of agricultural R&D. From some data gathered during recent years in the European fruit markets, we know that some retailers base their profit on high-quality, high-price fresh fruits and vegetables. This is attained by applying a very strong selection for 'extra-fancy' quality that causes a very high proportion of rejects. These are then sold in second-class more economic retail markets. This situation raises the question of safety and value to the consumers in these markets. Efforts to ensure a high quality of fresh fruits and vegetables are being made worldwide.

Causes of mechanical damage

Bruising appears as a result of impacts and compressions of the fruits against other fruits, parts of the trees, containers, parts of any grading and treatment machinery and on any uncushioned surfaces. Severity of damage to the fruit is primarily related to: (i) height of fall; (ii) initial velocity; (iii) number of impacts; (iv) type of impact surface and size; and (v) physical properties of the fruit, related or not to maturity.

Fruit that is marketed to be consumed fresh is harvested manually. This means that fruits are picked one by one by hand-pickers and placed into some type of containers, then transported to a packing-house in different types of vehicles (trucks, tractors plus trailers). There, fruits are subjected to a number of operations, which vary greatly between commodities, but which combine similar individual treatments. Table 9.1 shows a list of such operations, and a combination of these is made up for the different species; it is important to state that any combination of treatments may be applied to freshly picked fruits and also to stored fruit, at shorter or longer periods of time after harvest; also, the last operations, transportation to wholesale and retail, may add up to various cycles (two, three) as the product proceeds from production site (a different country in export operations) to retail market. This will create important differences in the damage that is caused to the fruits, due to the significant changes which may occur in their physical and physiological properties, related to variations in time lapses and environmental conditions. Also, transportation/storage/grading may have to be combined with a cooling chain; the maintenance of this whole system is of

Table 9.1. Harvesting and handling operations used in fruit marketing.

Harvest into:
  buckets
  field-boxes, or
  pallet boxes
Transportation to packing-house
Dumping, dry or into water
Washing
Waxing
Sorting
Sizing
Packing
Cooling
Storing
Transportation to:
  wholesale markets
  chain store distribution centres
  retail markets
Shelf storage

great significance in the changes of the mentioned fruit properties, and therefore in their susceptibility to damage at any stage.

It appears that most of the studies on damage caused to deciduous fruits have been carried out in relation to mechanical harvesting, as this is the main concern in the design and in the adoption of this type of equipment, for industry as well as for fresh market fruit. Diener and Fridley (1983) emphasize the importance of pruning to avoid excessive impacts of fruits against limbs during their fall through the tree, when harvested by the shake–catch method; canopy shapes are recommended. Other innovative procedures for catching the fruits without submitting them to drops have been investigated, as well as special padded-roller conveyors for minimizing impacts on the fruits during in-field handling. Today, no mechanical harvesting is used for fresh market fruits, but some of the conveying devices have been introduced in fruit grading machines. They consist mainly of padded rolls which feed the fruits instead of letting them fall between conveyors; they eliminate the drop-height and acceleration to the fruit, thereby eliminating impact.

Cushioning of the catching surface of a harvester is supposed to eliminate bruising. Zocca *et al.* (1985) report the results of shake–catch harvesting of two varieties of apples and one of peaches. The catching surface was covered by a well-cushioned PVC material. Thirty percent of the fruits were bruised in the case of peach variety Babygold 9; 50% and 63% in the case of apple varieties Abbondanza and Renetta Grigia. This result shows the different susceptibility to impact by different fruits (lower for peaches than for apples), and also the high impact damage applied to the fruits by this method of harvesting. Similar results have been obtained during recent years by other authors working with different varieties of apples (Bilanski and Menzies, 1984). Bennedsen (1986) used special 'X' and 'L' shaped foam shapes mounted on the catching frame. Laboratory tests indicate that only 8−11% of the apples dropped on this catching frame were damaged.

Bruising could be maintained at a level of 2−7% in Granny Smith apples when the orchard was 'H'-shape-canopy-trained and the cushioned catching frame could thus be positioned immediately below the fruits. Using the same canopy training in raspberries, machine-harvested berries were of excellent quality (no damage) when harvested with a rotating-finger harvester (Dunn *et al.*, 1976). Commercial tomato varieties may be dropped as high as one metre without damage.

These results show clearly that different species of fruits (and varieties within species) will react very differently to similar damaging inputs when harvested or handled. This leads to some conclusions on impact damage mechanisms, some of which will be studied further.

Recent studies show a renewed interest in introducing mechanical harvesting of fresh market Golden Delicious apples (Peterson and Miller, 1988). An impact shaker and a rod-press harvester were able to pick up to 95% of the fruits with an average of up to 90% extra fancy/fancy quality. Burton *et al.* (1989) made studies to quantify and identify the areas where damage occurs to apples during bin filling and handling, and during transport to the packing-house, with the aim of introducing improvements in the whole system. The importance of using well-damped suspensions on the trucks and a cushioning lining in the bins was emphasized for long-distance transportation (50−500 km). Relative to transportation, Schulte Pason *et al.* (1989) show the strong relationship between distance of travel, degree of apple bruising and number of impacts greater than 20 g as determined by an IS (Instrumented Sphere, also called SEP: Simulated Electronic Product). A recent development of these SEPs, impact data acquisition units, makes it possible to study fruit harvesting and handling equipment for the level of damage applied to the products, as well as the specific damage sites (Nissen and Kampp, 1987; Anderson, 1990).

Fruits may be classified into different types regarding their most evident physical properties, which are responsible for their susceptibility to bruising. Such a classification is very inaccurate, as many fruits may change from one type to another during ripening or when subjected to different conditions. Nevertheless, some groupings may be made.

Firstly, one may distinguish between 'rigid' ('hard') and 'liquid' ('plastic' or 'soft') fruits (Fig. 9.1).

- Rigid fruits are those whose strength is based on a mostly rigid structure, surrounded by a thin elastic skin: apples, pears, peaches, nectarines, apricots, avocados, mangoes, papaya, kiwi fruits, potatoes, etc. In this

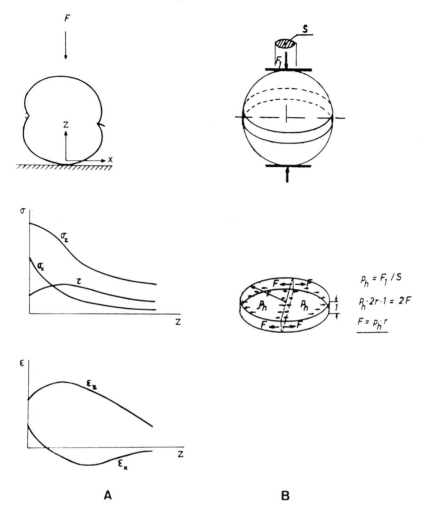

Fig. 9.1. Fruits can be considered as elastic solids (A) or as liquid-filled elastic spheres (B) (Ruiz-Altisent and Gil, 1979); these mechanical models are extreme approximations to real fruits (p_h = hydrostatic pressure; F = force exerted to the skin due to p_h; (see also Table 9.2).

type of commodity resistance is based in the fruit flesh, in its histological and physiological characteristics mainly.

- Liquid fruits are made up by a liquid or 'soft' mass contained in a mostly elastic skin, their resistance being based on this skin: examples include plums, tomatoes, grapes, cherries and berries. It is known that many rigid fruits gradually become soft as they mature.

- Mass of the fruits is crucial in bruising susceptibility, since impact energy is, as known, directly dependent on the dropping mass. Small fruits will be handled much more safely than large fruits.
- Thick-skinned fruits, like melons, water-melons and bananas are very resistant to impacts, but skin rupture problems may occur with these fruits.
- Fibrous fruits like pineapples react in a very different manner to impacts, and have not been widely studied.
- The stone in fruits is the cause of internal bruising for higher-energy impacts in some fruits and varieties.

All these different types of fruits will have to be studied accordingly when trying to describe and to model their behaviour.

External damage of the fruit skin can be caused by friction and abrasion against bin walls and conveyors. Oranges are especially susceptible to this type of damage (Chen and Squire, 1970; Juste *et al.*, 1990), as are other citrus fruits. Also, some pear varieties are very easily damaged by abrasion (Valenciano García, 1990). Peeling or 'scuffing' of potatoes and other produce has been studied, and some testing devices exist to measure susceptibility to this type of damage on the skin (Muir *et al.*, 1990).

Cuts and punctures represent severe damage, caused by inappropriate equipment or handling; they are not related directly to fruit properties, and can be avoided by proper care of the equipment and of the handling systems.

Measuring and modelling the contact phenomena

Contact models: applications

Various theoretical models have been used to explain and analyse the impact problem as applied to fruits. The first one was presented many years ago, and consists of considering a fruit as an elastic (generally spherical) body and applying the Hertz contact theory further developed by Shigley (Horsfield *et al.*, 1972; Rumsey and Fridley, 1977). This approach has been shown to be only approximately applicable, but has yielded much interesting information on many fruits, especially those we have called hard or rigid fruits. The elastic contact problem in fact describes the internal stresses and strains created in and below the contact area between fruit and impacter of elastic, rigid, isotropic and semi-infinite bodies. It states that bruising can initiate at a certain depth below the skin, where the maximum shear stresses and strains appear. Table 9.2 shows a summary of the most relevant mathematical relations of this model. At first it was mainly applied to peaches and pears, but later many applications have been and are being published on apples, and even plums, cherries, potatoes and many more (Chen *et al.*, 1984;

Table 9.2. Summary of the relevant features and mathematical relations of the elastic and of the viscoelastic models.

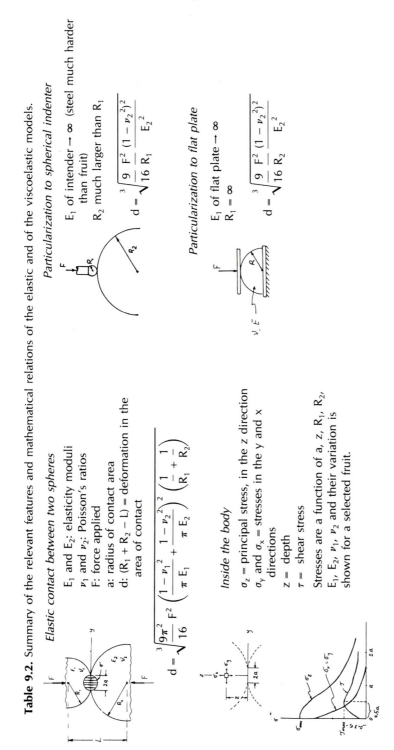

Elastic contact between two spheres

E_1 and E_2: elasticity moduli
ν_1 and ν_2: Poisson's ratios
F: force applied
a: radius of contact area
d: $(R_1 + R_2 - L)$ = deformation in the area of contact

$$d = \sqrt[3]{\frac{9\pi^2}{16} F^2 \left(\frac{1 - \nu_1^2}{\pi E_1} + \frac{1 - \nu_2^2}{\pi E_2} \right)^2 \left(\frac{1}{R_1} + \frac{1}{R_2} \right)}$$

Inside the body

σ_z = principal stress, in the z direction
σ_y and σ_x = stresses in the y and x directions
z = depth
τ = shear stress

Stresses are a function of a, z, R_1, R_2, E_1, E_2, ν_1, ν_2 and their variation is shown for a selected fruit.

Particularization to spherical indenter

E_1 of intender $\rightarrow \infty$ (steel much harder than fruit)
R_2 much larger than R_1

$$d = \sqrt[3]{\frac{9}{16} \frac{F^2}{R_1} \frac{(1 - \nu_2^2)^2}{E_2^2}}$$

Particularization to flat plate

E_1 of flat plate $\rightarrow \infty$
$R_1 = \infty$

$$d = \sqrt[3]{\frac{9}{16} \frac{F^2}{R_2} \frac{(1 - \nu_2^2)^2}{E_2^2}}$$

Table 9.2 *Continued*

Viscoelasticity

Combines elastic and viscous properties

Deformation rate affects stress–strain relationship.

$$E = \frac{\sigma}{\epsilon} = \text{modulus of elasticity}$$

$\dot{\epsilon}_1$ and $\dot{\epsilon}_2$: strain rates, $\dot{\epsilon}_1$ being lower than $\dot{\epsilon}_2$

Spring and dashpot represent the elastic and viscous parts of biological tissues.

The *Maxwell model* combines both elements:

$$\dot{\epsilon} = \frac{\dot{\sigma}}{E} + \frac{\sigma}{\eta}$$

and its force–deformation relationship fits to the one shown by biological tissues.

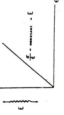

Generalized Maxwell model is the sum of many Maxwell models.

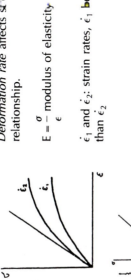

Stress-relaxation test

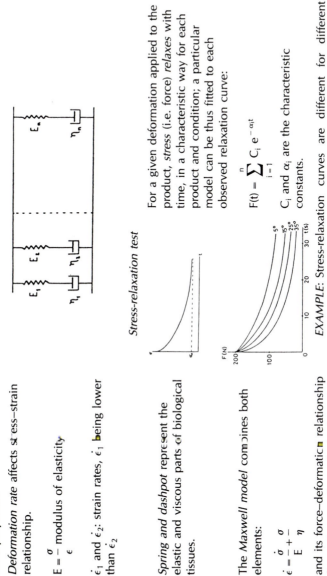

For a given deformation applied to the product, *stress* (i.e. force) *relaxes* with time, in a characteristic way for each product and condition; a particular model can be thus fitted to each observed relaxation curve:

$$F(t) = \sum_{i=1}^{n} C_i\, e^{-\alpha_i t}$$

C_i and α_i are the characteristic constants.

EXAMPLE: Stress-relaxation curves are different for different temperatures in apples (Gil *et al.*, 1984).

Hemmat and Murfitt, 1987; García Alonso *et al.*, 1988; Lichtensteiger *et al.*, 1988; Blahovec, 1990; Sinn, 1990).

When studying different kinds of fruits and fruit-probes, in different maturity and turgidity stages, only very few show distinct shear failure

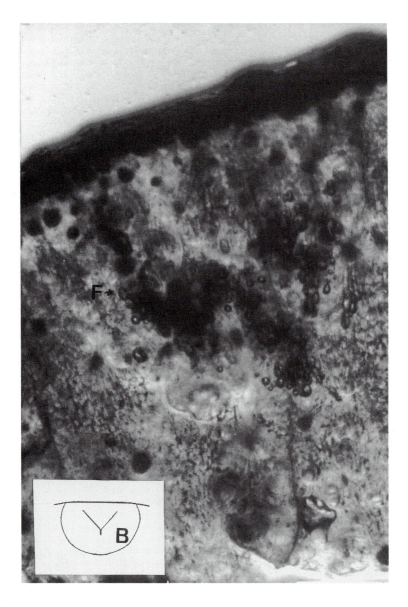

Fig. 9.2. Conical fracture (F) observed in pears (variety Limonera, 6 cm drop, 50 g sphere of 9 mm diameter); B: bruised tissue. (See also Fig. 9.7 (b).)

surfaces (conic, at 45° slope; see Fig. 9.2), and many show rather horizontal failure planes (especially apples: Jarimopas, 1984; Ruiz Altisent *et al.*, 1989; Fig. 9.3), other types of failure patterns, or no failure surfaces at all (Ruiz Altisent *et al.*, 1989).

Size of the observed bruised volume is not correlated to the calculated values of shear stress in many cases, especially when testing fruits at increasing maturity. This shows that the mechanical properties of fruits vary accordingly, and so do their impact responses; also, other physiological parameters of the fruit flesh are involved in the initiation and development of bruises (Rodriguez, 1988).

Impact parameters (especially mass, initial velocity, input energy, impulse, and radii of curvature) also affect greatly the impact phenomenon, as well as the macro- and microstructural properties of the fruits. Firmness of fruits is used as an index of maturity, and it is measured by puncture of the flesh. Discussion on the influence of firmness in bruising susceptibility of selected fruits is still open. On the basis of the elastic theory, a higher modulus of elasticity leads to higher stresses, and therefore higher bruising susceptibility should be observed; puncture firmness is generally correlated to modulus of elasticity. Common practice shows, however, that softer (less firm) fruits are usually more easily bruised during handling. In fact, puncture (Magness–Taylor) firmness is a complex test, combining different aspects of tissue strength, which vary in different ways with tissue properties.

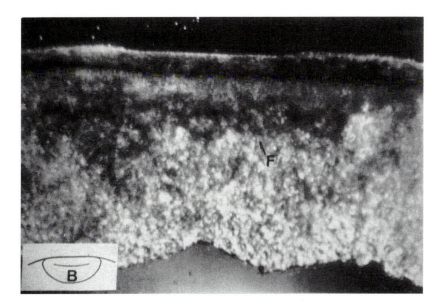

Fig. 9.3. Horizontal failure surface observed in apple bruises (Starking variety; B: bruised tissue; F: failure surface.)

Fruit tissue is made up by cells, their walls forming the rigid structure of the pulp; these cells are bonded together by a connecting substance, and the tissue contains varying proportions of gaseous spaces, free liquids or even oils (Pitt, 1982; Rodriguez *et al.*, 1990; Ruiz Altisent *et al.*, 1989; García Alonso *et al.*, 1988). Both cell walls and bonding material change greatly with ambient conditions and with ripening. Summing up, a fruit is in fact a physical body with continuously changing properties; and the response of fruits to contact loading is very much dependent on the type of loading.

Taking into account the viscous behaviour of the liquid fraction of fruit tissue, contact has also been modelled using viscoelastic models. Viscoelasticity has been extensively studied and applied to static testing of biological products (Rumsey and Fridley, 1977; Gil *et al.*, 1984; Gil, 1990). Hamann (1970) applied the viscoelastic model to the impact case, showing that stresses are distributed in the impacted tissue somewhat differently than when calculated by pure elasticity, but the applications shown do not significantly improve the elastic solution. Duration of impact is very low for fruits (5–8 ms; see below), whereas viscoelastic time constants are much longer (minutes). Rumsey and Fridley (1977) developed a finite-element analysis of contact stresses for elastic as well as for viscoelastic spherical bodies in contact, subjected to static load as well as to impact. By comparison of the distribution of stresses calculated by the analytical versus the finite-element procedures they concluded that for elastic homogeneous bodies no differences appear for both solutions. However, when material properties vary within the body, the finite-element method is the most appropriate for calculating internal stresses caused by static or impact loading (Fig. 9.4).

Gil *et al.* (1988) studied the effect of temperature and firmness differences on the viscoelastic modelling of the static stress-relaxation test of cylindrical probes of apple and pear flesh. Relaxation was significantly faster and larger for warmer and for less-firm samples, and size of bruises was observed to be smaller when impacting the corresponding fruits.

All these theoretical approaches for calculating internal stresses as a result of static contact and of impact, are only applicable for very small strains. Their application to problems where large strains occur, as is the case for most agricultural products, becomes questionable. Nevertheless, the description of the theoretical distribution of stresses and strains as a result of loading yields very useful information to be compared with empirical observations.

Testing devices

Testing devices which have been developed in recent years to apply and analyse controlled impacts to fruits include instrumented pendulums (Holt and Schoorl, 1984; Hughes and Grant, 1987), free-falling instrumented devices (Chen *et al.*, 1985), and spring-actuated falling rods (Gahtow, 1990).

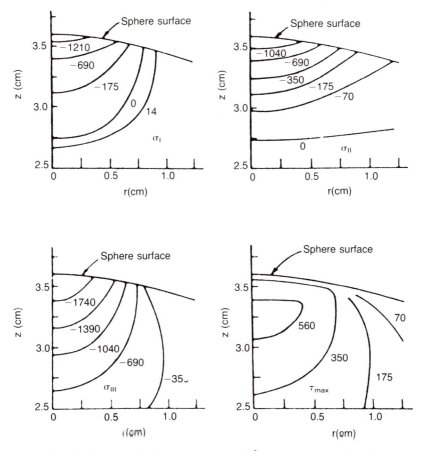

Fig. 9.4. Principal stresses (kPa) at time 16×10^{-3} seconds after a 5 cm drop as predicted by finite-element modelling of a viscoelastic sphere. The stress σ_1 is the maximum principal stress and σ_{III} is the minimum principal stress, so that $\tau_{max} = (\sigma_1 - \sigma_{III})/2$. (Rumsey and Fridley, 1977.)

Chen *et al.* (1987) and Garcia *et al.* (1988) used an impact testing device, consisting of a free-falling impacting rod with a changeable spherical tip, instrumented with a miniature accelerometer. The response of the sample to the applied impact can be fully characterized by the force/time curve obtained by each impact. Analysis of these impact curves has been the basis for various impact studies in fruits. Significant results were obtained, relative to the parameters which best characterize the impact response of these materials and to their correlation to fruit damage, variety and ripeness level of the fruits (Ruiz Altisent, 1990). Bruise damage, measured as the size and/or the volume of the affected fruit tissue is primarily related to input energy (that is, drop height) for a given variety in a given ripeness stage and physical

Table 9.3. Summary of impact parameters relevant to characterize impact response in fruits. First group are energy related parameters; second group are both related to energy of impact and to fruit texture; third group are significantly related to ripeness of fruits. (Selected apple and pear varieties; García Alonso *et al.*, 1988.)

Name of parameter	SI units	Symbol
Maximum deformation	mm	DM
Permanent deformation	mm	DP
Critical depth (maximum shear stress location Hertz model)	mm	PC
Maximum mechanical impulse	Ns	IM
Maximum bruise depth	mm	PM
Maximum bruise width	mm	AM
Percentage of rebound energy	%	%E
Maximum impact force	N	FM
Optimum slope force/time	N/s	F/T
Calculated coefficient	N^2/s	IF × F/T
Rebound velocity	m/s	VF
Total impact duration	ms	T
Final impact duration	ms	TF
Time to maximum force	ms	TM
Increment TT–TF	ms	IT
Optimum slope force/deformation	N/m	FD
Apparent dynamic modulus of elasticity	Pa	ME
Maximum shear stress	Pa	EC

condition. The relevant impact response parameters are maximum deformation (DM), permanent deformation (DP), maximum impulse (IM), maximum impact force (FM) and impact duration (T). A correspondence factorial analysis is applied, to group the most significant parameters on the basis of their linear relation to: (i) input energy; or (ii) ripeness level. Table 9.3 shows all the relevant parameters and the groupings resulting from the described tests.

With this empirical approach of impact study and analysis, parameters of different origin can be combined and analysed jointly: (i) measured parameters; (ii) parameters calculated from the acceleration data; and (iii) parameters calculated from the theoretical models. In tests carried out on four varieties of Asian pears (Chen *et al.*, 1987), significant effects of variety, time in cold storage, and time of ripening in 20 °C room, were observed on impact and compression bruise sizes, and on most of the impact parameters. Correlations were established between bruise dimensions and some impact response parameters. Firmness appeared always negatively correlated to bruise dimensions.

Rodriguez and Ruiz (1989) applied linear regression models to the impact response parameters in pears of the Blanquilla variety, trying to explain bruise size, defined by its depth and diameter. The included parameters were: maximum impact force, maximum impact deformation, permanent deformation, input energy, absorbed energy, impact duration, firmness, acidity, soluble solids and the soluble solids / acidity ratio. Around 57% of the total variation could be explained by these parameters, being most of the variation explained by input energy alone; when fruits were in the senescent stage (ripe to overripe), deformations (maximum and permanent) were the most relevant parameters in explaining bruise size. Pears change very significantly from rigid to plastic with advancing maturity.

Bruise volume has been used by various researchers rather than depth and diameter to evaluate bruise severity; Kampp and Nissen (1990) studied the susceptibility to impact, applied by an instrumented pendulum, of seven varieties of apples. As in the results reported by many other researchers, high correlations were found between input energy (E_{abs}) and bruise volume (V) for any of the three (early, mid- and late harvest) samples of each variety. The important result was, however, that expressing the impact susceptibility as the regression coefficient a in the expression:

$$V = aE_{abs} + b \quad \text{(a in ml/J; b in ml)} \quad [9.1]$$

The impact susceptibility is different for every sample of every variety. From this and many other similar results it is concluded that neither input nor

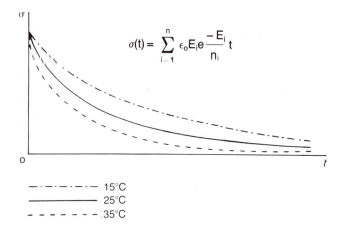

$$\sigma(t) = \sum_{i=1}^{n} \epsilon_0 E_i e^{\frac{-E_i}{n_i} t}$$

———·———·— 15°C
———————— 25°C
— — — — — — — 35°C

Fig. 9.5. Stress-relaxation curves of cylindrical probes of apple flesh, at different temperatures, for equal initial stress values (Gil *et al.*, 1988; Gil, 1990). Stress-relaxation was observed to be faster and larger for warmer apple tissue. The variation of stress with time at constant deformation is modelled by a three- or four-elements Maxwell model, where ϵ_0 = initial strain, E_i = elastic constants and n_i = viscous constants.

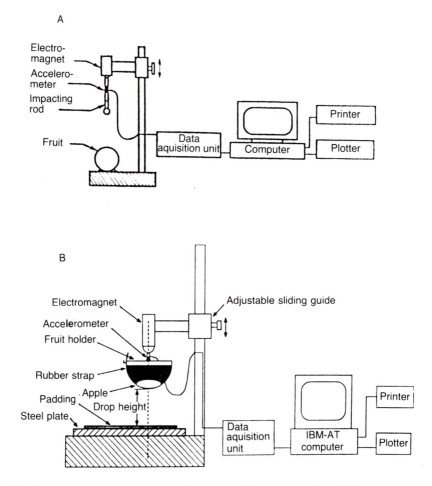

Fig. 9.6. Impact testing devices and associated instrumentation: (A) Dropping steel sphere on fruit; (B) dropping fruit on steel plate and padding materials. (Chen *et al.*, 1985; Chen and Yazdani, 1989.)

absorbed energy are in themselves sufficient to predict impact sensitivity of apples and consequently of other fruits. Other parameters related to: (i) the impact process; (ii) the response of the fruit due to its physical properties; and (iii) the structure and physiology of the fruit, must be included to explain bruising.

Free fall of instrumented fruits was used by Chen and Yazdani (1989; see Fig. 9.6). The degree of bruising of Golden Delicious apples dropped from different heights onto different impacting surfaces (padded differently) could be predicted by multiple regression models based on: (i) measured and calculated impact parameters; and (ii) Fourier-Transform coefficients of the impact acceleration curves. The relevant parameters were maximum value of force/time rate change (F/T slope), maximum deformation (DM), and absorbed energy (EAB); further, maximum force (FM) and duration or time of impact (T) were significant in the regression equation. These results are basically coincident with those obtained by other researchers already mentioned.

Siyami *et al.* (1988) used an impact table to perform free fall tests on apples. Application of the elastic contact model was used to develop an equation for predicting bruise diameter: BD = Function (W = apple weight, H = drop height, D = apple diameter and F = Magness–Taylor firmness). A multiple linear regression model, based on the parameters apple diameter, Magness–Taylor force, maximum acceleration, and total velocity change (v–v rebound), could accurately predict bruise diameter in Ida Red apples. The theoretical solutions of the elastic model, as well as a similar one developed including plastic deformation, both underestimated bruise diameter in these fruits. These results show again that, using the appropriate measuring instrumentation, empirical models based on measured and, eventually, calculated parameters are appropriate for studying impact response and for predicting bruise size.

Lichtensteiger *et al.* (1988) used a drop-testing apparatus where the samples were released from specific heights onto a rigid aluminium plate instrumented with a force transducer. Various types of models (fabricated balls) and red tomatoes were tested. Changing the properties of the shell in relation to the internal material of the tested balls showed that the shell effect is prevalent when the internal structure is softer than a relatively thin shell. When the internal material is stiffer than the shell, no shell effect was observed. This result shows that the effect of the skin when testing hard or soft fruits is in fact relevant in the response to impacts, and it will be different for changing ripeness of the fruits.

Brusewitz and Bartsch (1989) also used dropping of fruits (five varieties of apples) onto a plate, instrumented with a piezoelectric force transducer. They showed that the relation 'bruise volume / absorbed energy' changes gradually with storage time, decreasing, and consequently also with firmness. Other references show different or opposite results (Hung and Prussia, 1988; Holt and Schoorl, 1984), namely an increase in 'bruise volume / absorbed energy' when reducing firmness. They also found that impact contact time (T) was closely correlated with decreasing firmness, as well as the relation impact force / contact time (FM/T), very much in agreement with all the impact parameters research results found so far.

Timm *et al.* (1989) studied impact susceptibility in apples. Fruits were dropped onto an impact surface, carrying an accelerometer on the opposite side. Multiple linear regression based on similar parameters as the ones reported by Siyami *et al.* (1988) were very good predictors for bruise dimensions, on both hard and padded surfaces.

Sinn (1990) performed free-fall impact testing of cherries and plums. The stress–strain behaviour of soft fruits is not like the one shown by hard fruits, but a good correlation was observed between impact force and fruit damage. Fruits were dropped onto a plate instrumented with a force transducer. Data were gathered on the relative effect of different padding materials on the maximum force (FM) determined at impact.

Bruising can also be caused by static and quasi-static contact loading. All the mechanical models which have been applied to describe impact were developed for static contact, with similar results and with the same restrictions for accuracy. Viscoelasticity becomes important in static loading. Its main effect is that instantaneously applied stress relaxes with time, and instantaneous strain 'creeps' with time; the resulting stresses are therefore lower than in impact, for similar energy inputs, and the strains larger. Chen *et al.* (1987, cited above) compared the bruises produced by compression and by impact, and they observed that bruise pattern could be very different in both loading speeds: for a variety of pears, long spikes extended radially from the impact area into the fruit, showing that loading rate has a great influence when analysing strength and failure of fruits. Results obtained so far indicate that higher loading rates and higher firmness (hardness) of fruits usually show shear failure patterns; slower loading rates and lower firmness show normal stress or strain failure (Chen and Sun, 1984, Ruiz Altisent *et al.*, 1989). Therefore, discrepancies found in the results of different researchers in relation to fruit tissue failure should be due to these mentioned differences in loading rate and fruit properties in their testing procedures. This refers also to the bruise volume / energy ratio discrepancies, discussed earlier.

In electron microscope studies carried out in a variety of apples (see below), it was observed that degradation of cells was different when impacted than when slowly compressed.

Structure of bruises

Closer observation of bruises caused by impact shows that different species of fruits (Rodriguez, 1988; Ruiz Altisent *et al.*, 1989; Rodriguez *et al.*, 1990, studying apple and pear fruits) show different bruise sizes and patterns which appear in the absence of any significant variation in other relevant parameters (like radii of curvature of the fruits or the impacters, or the energy of impact). Therefore, they have to be related to structural differences between these fruits, which in fact are very important. These include size and shape of cells in hypodermis (first cell layers below the skin) and pulp, and the presence of intercellular spaces. Table 9.4 includes the most important differences in the structure of the fruit tissues, gathered for apples and pears. Figure 9.7 shows different models of average bruises observed in the varieties studied. The differences between them affect the following.

Table 9.4. Differences in tissue structure between apples and pears. (Ruiz-Altisent *et al.*, 1989.)

Apple	Pear
Epidermis (skin)	
Squared cells	Flat cells
Many void spaces	No void spaces
Suberized cells	
Thin-walled hairs	Thick-walled hairs
Hypodermis (below skin)	
Isodiametric cells of variable size	Polygonal cells with thick walls
Large intercellular spaces	Few, small intercellular spaces
Parenchima (flesh)	
Large, irregular cells	Polyhedric cells of smaller size
More and larger intercellular spaces	Esclereids: thick-walled cells in groups
Many fibres	Calcium oxalate (hard) crystals
Lower apparent density	Higher apparent density

1. The relation depth/width, significantly higher in pears than in apples. For 12 cm drops (0.06 J), the average depth is similar for both apple varieties (Golden Delicious and Starking) and for two of the pear varieties (Decana and Blanquilla) – that is, 2.5–3.5 mm, with no significant change for an 11-week period of cold storage; however, bruise width is significantly larger (7–8 mm) for both varieties of apples (pears 3–5 mm). The larger and more numerous intercellular spaces present in apple tissue cause transversal failure surfaces, which are able to absorb much of the strain during impacter penetration, thus dissipating much of the stresses and preventing the formation of deep bruises.

2. The size, shape and number of fractures observed inside the bruised tissue. Close observation of the bruises makes it possible to identify the presence of discontinuities, eventually fractures, in what may be the maximum stress/strain location in the bruised area when impacted. They always appear some millimetres below the skin, and centred in the bruise itself. In the case of apple varieties, somewhat curved horizontal fissures appear (see also Jarimopas, 1984), not completely separated but rather showing cell walls like bridges between both sides; the cells are crushed together, with no evidence of having slid, so they may not be properly considered fractures. On the other hand, in the case of Decana pears, real conical and vertical fractures are observed (Fig. 9.8), suggesting stress failure as described for some fragile materials. This type of fracture has been described previously in Asian pears (Chen *et al.*, 1987).

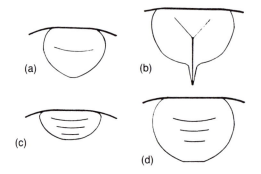

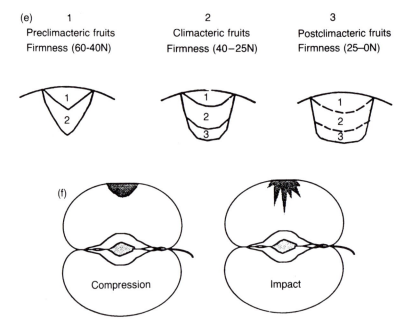

Fig. 9.7. Types of bruises and fractures or fissures observed in varieties of apples and pears (12 cm drops): (a) Limonera pear; (b) Decana pear; (c) Golden Delicious and Starking apples; (d) and (e) Blanquilla pear (1 hard, 2 ripe, 3 soft); (f) comparison of compression and impact bruise patterns in Asian pears.

3. For low-energy impacts, although a bruise appears (of smaller size) no discontinuity is observed.

4. It is important to note that after some time, the totality of the stressed tissue becomes discoloured, not only the ruptured spot or surface. It is also observed that the specific reactives for polyphenoloxidase activity become

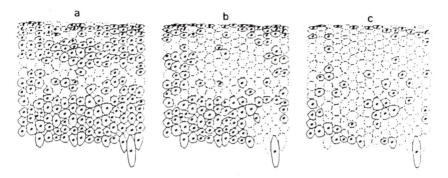

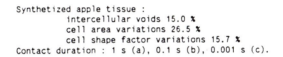

Synthetized apple tissue :
        intercellular voids 15.0 %
        cell area variations 26.5 %
        cell shape factor variations 15.7 %
Contact duration : 1 s (a), 0.1 s (b), 0.001 s (c).

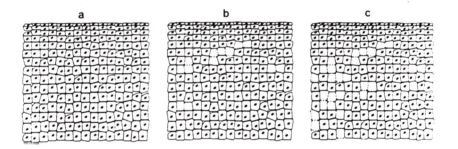

Synthetized apple tissue :
        intercellular voids 0.1 %
        cell area variations 9.7 %
        cell shape factor variations 7.0 %
Contact duration : 0.1 s (a), 0.0001 s (b), 0.00001 s (c).

Fig. 9.8. Mathematical graphical representation of apple tissues of different structure. Simulation of contact durations: 1 to 10^{-3} seconds for the 'soft' tissue (top); 0.1 to 10^{-5} seconds for the 'packed', hard tissue (bottom). (Roudot and Duprat, 1990.)

active in the total affected area (Rodriguez, 1988), very clearly delimiting the tissue that has been damaged.

Rodriguez *et al.* (1990) made transmission electron-microscope studies in Granny Smith apples, showing that after a few hours, cells of the bruised area are altered, with intensive vesiculation, either in the vacuole (inside the cell) or in the middle lamella region between adjacent cells. It was observed that cell wall rupture is not necessary to initiate bruise reactions; if it occurs, it may additionally leak the altered compounds out of the cells to the intercellular spaces. Thus, the browning reaction in fruit under applied loads can take place either outside or inside the cell. Low stresses applied to tissue cells which cause no rupture of cell walls, cause bruising, developed internally in the cells.

Roudot and Duprat (1990) used a graphical model to simulate the effect of these structural differences on failure in apple flesh. The model starts considering that apple flesh is not homogeneous and that it contains a variable proportion of intercellular spaces. With a special two-dimensional tessellation model of the cellular tissue it was possible to simulate the collapse of cells as a result of varying levels of loading, time of application (1, 0.1 or 0.001 seconds), percentage of intercellular voids and cell-covered surface, and of a factor describing the shape of the cells. Very different failure patterns could be obtained and compared to real bruises. Interesting observations were made, showing that collapsed cells increase with void space percentage and hetereogeneity in cell size and shape increases.

Biological variables

As has been stated already, the properties of fruits change in relation to many biological variables. These influences begin in the early stages of development of the fruit on the plant, caused by variations in varietal, agronomic and climatic conditions, and continue throughout the whole growing season. Some attempts have been carried out to explain the influence of such variables in the susceptibility of fruits to mechanical damage.

Johnson and Dover (1990) studied some factors influencing the bruising susceptibility of a variety of apples (Bramley's Seedling). Fruits from 24 commercial orchards were tested during six seasons. It was observed that bruising susceptibility (measured by means of an instrumented pendulum, applying 0.19 J of impact input energy) varies in a greater measure within a season (between orchards) than between seasons. There was no evidence of any agronomic or microclimatic factors which might be responsible for such differences – neither soil management systems (complete grass covering, strip-herbicide or overall herbicide), nor nitrogen fertilizer applications or mineral composition of leaves or fruits. Bruise volume was negatively correlated to fruit firmness (Magness–Taylor puncture test). Also, larger fruits were observed to be more susceptible to bruising, within samples of the same orchard; cells of such fruits are larger as well as their intercellular spaces. Water loss appeared to increase bruising resistance to the fruits. Bruise volume increased significantly with picking date, as shown in Fig. 9.9.

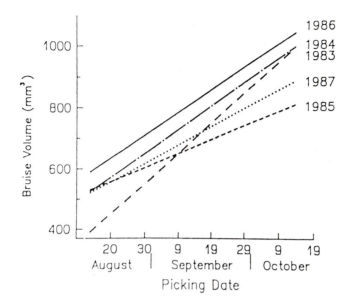

Fig. 9.9. Regression lines relating bruise volume to harvest date for Bramley's Seedling apples (impact energy 0.19 J). (Johnson and Dover, 1990.)

Hatfield and Knee (1988), Pitt (1982), Pitt and Davis (1984) (cited in Johnson and Dover) and other authors have associated turgor pressure with a reduction in the strength of apple tissue. Water loss appeared to increase bruising resistance of the fruits studied by the cited authors.

Jaron and Recasens (1990) tested the effect of calcium treatments on the physical properties of apples. Many researchers have observed an increase of tissue firmness when treated with calcium. This element contributes to maintaining cell membrane and cell wall integrity. Samples of Golden Delicious apples were treated with different calcium solutions, weeks before harvest. Static and impact tests were applied to the fruits, after harvest and after various intervals of cold storage. Significant differences were observed in the firmness of fruits subjected to some of the calcium treatments, as well as a reduction in bruising susceptibility.

References

Anderson, G. (1990) The development of artificial fruits and vegetables. *Proceedings of the European Workshop on Impact Damage of Fruits and Vegetables (EWIDF)*, Zaragoza (Spain), 27–29 March. FIMA, 90, pp. 133–9.

Bellon, V. (1990) Feasibility of new types of non-contact evaluation of internal quality inspection by spectroscopic techniques: near infrared, nuclear magnetic resonance . . . *Proceedings of the European Workshop on Impact Damage of Fruits and Vegetables (EWIDF)*, 27–29 March, Zaragoza (Spain). FIMA, 90, 143–53.

Bennedsen, B. S. (1986) Damage reduction in mechanical apple-harvesting. ASAE paper 86–1071, ASAE, St Joseph, MI, USA.

Bilanski, W. K. and Menzies, D. R. (1984) Bruising related to mechanical harvesting of apples and peaches. *Proceedings of the International Symposium on Fruit, Nut and Vegetable Harvesting Mechanization*, Bet Dagan, Israel, 5–12 October 1983, 376–81.

Blahovec, J. (1990) Impact damage in potato tubers. *Proceedings of the European Workshop on Impact Damage of Fruits and Vegetables (EWIDF)*, Zaragoza (Spain), 27–29 March, FIMA, 90, 81–5.

Burton, C. L., Brown, G. K., Schulte Pason, N. L. and Timm, E. J. (1989) Apple bruising related to picking and hauling impacts. ASAE paper 89-6049, ASAE, St Joseph, MI, USA.

Brusewitz, G. H. and Bartsch, J. A. (1989) Impact parameters related to postharvest bruising of apples. *Transactions of the ASAE* 32(3), 953–7.

Chen, P. and Squire, E. F. (1970) An evaluation of the coefficient of friction and abrasion damage of oranges on various surfaces. *Transactions of the ASAE* 14(6), 1092–4.

Chen, P. and Sun, Z. (1984) Critical strain failure criterion: Pros and cons. *Transactions of the ASAE* 27(1), 278–81.

Chen, P. and Yazdani, R. (1989) Prediction of apple bruising due to impact on different surfaces. ASAE paper 89-6501, ASAE, St Joseph, MI, USA.

Chen, P., Tang, S. and Chen, S. (1985) Instrument for testing the response of fruits to impact. ASAE paper 85-3537, ASAE, St Joseph, MI, USA.

Chen, P., Ruiz, M., Lu, F. and Kader, A. (1987) Study of impact and compression damage on Asian pears. *Transactions of the ASAE* 30(4), 1193–7.

Chen, S., Chen, P. and Herrmann, L.r. (1984) Analysis of stresses in fruit during an impact. ASAE paper 84-3554, ASAE, St Joseph, MI, USA.

Diehl, K. C. and Ordoñez, G. P. 1982. Normal strain failure due to contacts stresses. ASAE paper 82-3558, ASAE, St Joseph, MI, USA.

Diener, R. G. and Fridley, R. B. (1983) Collection by Catching. In: O'Brien, M., Cargill, B. F. and Fridley, R. B. (eds), *Harvesting and Handling Fruits and Nuts*. AVI, Westport, CT, USA, 245–99.

Dunn, J. S. and Stolp, M. (1981) Apple production for mechanized harvesting on the Lincoln Canopy System. *New Zealand Agricultural Engineering Institute, Extension Bulletin 10*.

Dunn, J. S., Stolp, M. and Lindsay, G. G. (1976) Mechanical raspberry harvesting and the Lincoln canopy system. ASAE paper 76-1543.

Fridley, R. B., Bradley, R. A., Rumsey, J. W. and Adrian, P. A. (1968) Some aspects of elastic behaviour of selected fruits. *Transactions of the ASAE* 11(1), 46–9.

Gahtow, R. (1990) A high-speed device for impact damage in fruits and vegetables. *Proceedings of the European Workshop on Impact Damage of Fruits and Vegetables (EWIDF)*, Zaragoza (Spain), 27–29 March. FIMA, 90, 49–58.

García Alonso, C., Ruiz Altisent, M. and Chen, P. (1988) Impact parameters related to bruising in selected fruits. ASAE paper 88-6027, ASAE, St Joseph, MI, USA.

Gil, J. (1990) Computer aided determination of mechanical parameters of fruits subjected to elastic impact and to stress relaxation. *Proceedings of the European Workshop on Impact Damage of Fruits and Vegetables (EWIDF)*, Zaragoza (Spain), 27–29 March. FIMA, 90, 45–8.

Gil, J., Ruiz Altisent M. and Chen, P. (1984) Numerically calculated viscoelastic constants related to bruising resistance. ASAE paper 84-6502, ASAE, St Joseph, MI, USA.

Gil, J., Ruiz Altisent M. and Camps, M. (1988). Influencia de la temperatura y de la madurez sobre las propiedades mecánicas de algunas variedades de pera y manzana. (Influence of temperature and maturity on some mechanical properties of selected varieties of pears and apples.) *Proceedings of III Congress of the Spanish Society of Horticultural Sciences*, Tenerife, 536–41.

Hamann, D. D. (1970) Analysis of stress during impact of fruit considered to be viscoelastic. *Transactions of the ASAE* 13(6), 893–900.

Hatfield, S. G. S. and Knee, M. (1988) Effects of water loss on apples in storage. *International Journal of Food Science and Technology* 23, 575–83.

Hemmat, A. and Murfitt, R. F. A. (1987) Maximum normal strain and internal bruising resulting from the impact loading of potato tubers. *EC Workshop on Potato Damage*, Kolding (Denmark), 14–15 October.

Holt, J. E. and Schoorl, D. (1977) Bruising and energy dissipation in apples. *Journal of Texture Studies* 7, 421–32.

Holt, J. E. and Schoorl, D. (1984) Mechanical properties and texture in stored apples. *Journal of Texture Studies* 15, 377–94.

Horsfield, B. C., Fridley, R. B., and Claypool, L. L. (1972) Application of the theory of elasticity to the design of fruit harvesting and handling for minimum bruising. *Transactions of the ASAE* 15(4), 746–50, 753.

Hughes, J. C. and Grant, A. (1987) A portable pendulum for testing dynamic tissue failure susceptibility of potatoes. *EC Workshop on Potato Damage*, Kolding (Denmark), 14 15 October

Hung, Y. C. and Prussia, S. E. (1988) Determining bruise susceptibility of peaches. ASAE paper 88-6026, ASAE, St Joseph, MI, USA.

Jaren, C. and Recasens, I. (1990) Testing the effect of calcium treatment of the physical properties of apples. *Proceedings of the European Workshop on Impact Damage of Fruits and Vegetables (EWIDF)*, Zaragoza (Spain), 27–29 March. FIMA, 90, 117–22.

Jarimopas, B. (1984) 'Failure of apples under dynamic loading.' Research Thesis. Israel Institute of Technology, Haifa.

Johnson, D. S. and Dover, C. J. (1990) Factors influencing the bruise susceptibility of Bramley's Seedling apples. *Proceedings of the European Workshop on Impact Damage of Fruits and Vegetables (EWIDF)*, Zaragoza (Spain), 27–29 March. FIMA, 90.

Juste, F., Fornes, I. and Castillo, S. (1990) Compression damage on citrus fruits. *Proceedings of the European Workshop on Impact Damage of Fruits and Vegetables (EWIDF)*, Zaragoza (Spain), 27–29 March. FIMA, 90, 105–12.

Kampp, J. and Nissen, G. (1990) Impact damage susceptibility of Danish apples. *Proceedings of the European Workshop on Impact Damage of Fruits and*

Vegetables (EWIDF), Zaragoza (Spain), 27–29 March. FIMA, 90, 97–102.

Kampp, J. and Pedersen, J. (1990) Quality of imported and domestic fruits and vegetables in the Danish retail trade with special reference to mechanical damages. *Proceedings of the European Workshop on Impact Damage of Fruits and Vegetables (EWIDF)*, Zaragoza (Spain), 27–29 March. FIMA, 90, 11–16.

Lichtensteiger, M. J., Holmes, R. G., Hamdy, M. Y. and Blaisdell, J. L. (1988) Impact parameters of spherical viscoelastic objects and tomatoes. *Transactions of the ASAE* 31(2), 595–602.

Muir, A. Y., Ostby, P. B. and Zender, F. N. (1990) Experiments with a mechanical skin scuffing device. Scottish Centre of Agricultural Engineering (SCAE), UK, Departmental Note No. 30.

Nissen, G. and Kampp, J. (1987) Bioteknisk Institut's 'Electronic potato'; *EC Workshop on Potato Damage*, Kolding, Denmark, 14–15 October.

Petersen, D. L. and Miller, S. S. (1988) Advances in mechanical harvesting of fresh market quality apples. *Journal of Agricultural Engineering Research* 42, 43–5.

Pitt, R. E. (1982) Models for the rheology and statistical strength of uniformly stressed vegetative tissue. *Transactions of the ASAE* 1776–84.

Pitt, R. E. and Davis, D. C. (1984) Finite element analysis of fluid-filled cell response to external loading. *Transactions of the ASAE* 27(6), 1976–83.

Rodriguez, L. (1988) 'Reacción a la magulladura producida por impacto en frutos.' (Bruising as a result of impact in fruits: Physiological and histological study, and evaluation techniques). Ph.D. Thesis, Polytechnic University Madrid. ETSI Agrónomos.

Rodriguez, L. and Ruiz Altisent, M. (in press) Bruise development and fruit response of pear cv. Blanquilla under impact conditions. *Journal of Food Engineering*.

Rodriguez, L., Ruiz, M. and DeFelipe M. R. (1990) Differences in the structural response of 'Granny-Smith' apples under mechanical impact and compression. *Journal of Texture Studies* 21, 155–64.

Roudot, A.-C. and Duprat, F. (1990) A graphical model to simulate the mechanical behavior of apple flesh. *Proceedings of the European Workshop on Impact Damage of Fruits and Vegetables (EWIDF)*, Zaragoza (Spain), 27–29 March. FIMA, 90, 35–42.

Ruiz Altisent, M. (1990a) Impact parameters in relation to bruising and other fruit properties. *Proceedings of the European Workshop on Impact Damage of Fruits and Vegetables (EWIDF)*, Zaragoza (Spain), 27–29 March. FIMA, 90, 27–33.

Ruiz Altisent, M. (1990b) Impact damage of selected varieties of apples, pears and other fruits. Research Note. *Proceedings of the European Workshop on Impact Damage of Fruits and Vegetables (EWIDF)*, Zaragoza (Spain), 27–29 March. FIMA, 90, 113–16.

Ruiz Altisent, M. and Matthews, J. (1990) Impact damage in fruits and vegetables. *Final Report* of the European Workshop on Impact Damage of Fruits and Vegetables (EWIDF). FIMA, Zaragoza (Spain), 28–29 March.

Ruiz Altisent, M., García, C. and Rodriguez, L. (1989) Impact bruises in Pomaceae fruits: Evaluation methods and structural features. *Proceedings of the IVth International Congress on Physical Properties of Agricultural Materials*, Rostock, Germany, 698–703.

Rumsey, T. R. and Fridley, R. B. (1977) Analysis of viscoelastic contact stresses in agricultural products using a finite-element method. *Transactions of the ASAE* 21(3), 594–600.

Schulte Pason, N. L., Timm, E. J., Brown, G. K., Marshall, D. E. and Burton, C. L. (1989) Apple damage assessment during intrastate transportation. ASAE paper 89-6051, ASAE, St Joseph, MI, USA.

Sinn, H. (1990) Mechanics of impact of cherries and plums falling on different kinds of catching surfaces. *Proceedings of the European Workshop on Impact Damage of Fruits and Vegetables (EWIDF)*, Zaragoza (Spain), 27–29 March. FIMA, 90, 69–77.

Siyami, S., Brown, G. K., Burgess, G. J., Gerrish, J. B., Tennes, B. R., Burton, C. L. and Zapp, R. H. (1988) Apple impact bruise prediction models. *Transactions of the ASAE* 31 (4), 1038–46.

Timm, E. J., Schulte Pason, N. L., Brown, G. K., and Burton, C. L. (1989) Apple impact surface effects on bruise size. ASAE paper 89-6048, ASAE, St Joseph, MI, USA.

Valenciano García, J. (1990) Estimate of quality losses of fruit and vegetables in packinghouses with special reference to mechanical damages. *Proceedings of the EWIDF*, Zaragoza (Spain), 27–29 March. FIMA, 90, 19–25.

Zocca, A., Rosati, P. and Mazza, A. (1985) Performance of a new tree fruit harvester. ASAE paper 85-1565, ASAE, St Joseph, MI, USA.

Chapter 10

The Physics and Engineering of the Extraction of Protein from Green Crops

R. Řezníček

Introduction: The importance of protein separation from green crops

The extraction of part of the juice from plant tissues by the application of mechanical force (usually by pressing), possibly followed by the processing of the juice into protein–vitamin concentrate and of the cake into forage, is referred to as green crop fractionation (GCF). Using this procedure, farmers can obtain simultaneously high-fibre fodder for ruminants (from the cake) and a protein and vitamin fraction (from the juice) which is of high quality and high efficiency. The latter can be used as feed for monogastric livestock and can also serve for the production of protein–vitamin concentrates to be added to human food. There are projects using this technology in underdeveloped countries to alleviate the lack of food.

The bibliography in this and related fields is extensive, and the number of published papers and books on these problems is as high as 100,000. The first studies dealing with the operation-scale use of this technology appeared in the 1930s. In the subsequent period, the most important contribution to the development of the theory and practice of green crop fractionation has been made by Professor N. W. Pirie of the Rothamsted Experimental Station in the UK. Together with Professor S. Matai of the Indian Statistical Institute, Calcutta, he has published the *Leaf Protein Newsletter* for many years. Pirie (1978) is also the author of one of the first publications to deal, in a summarizing way, with green crop fractionation.

A number of international conferences on green crop fractionation have been held. The last one to date, held in Italy on October 1–7, 1989, was organized jointly by the Universities of Pisa, Perugia and Viterbo under the leadership of Professor C. Galoppini and entitled 'The Third International Conference on Leaf Protein Research'. The Third International Conference on the Physical Properties of Agricultural Materials, held in Prague in 1985, included one symposium on GCF. The Proceedings of this conference (Řezníček, 1988a) contain numbers of papers on GCF. Experiments with this technology have been conducted in almost all technologically advanced countries, aiming to utilize more effectively the nutrients contained in green plants and to reduce, at the same time, the specific power consumption of forage drying facilities. Experimental or operation-scale machine lines for green crop fractionation have been developed in at least ten countries. Detailed information on these facilities can be found in the monograph by Blahovec and Řezníček (1980). The survey of pertinent literature from all over the world, as given in that monograph, is also a good source of information. The latest information and findings are available in the *Leaf Protein Newsletter* mentioned above.

Findings presented in this chapter were gathered and written in 1990, taking into account research and practical results achieved all over the world. They are built on the basis of experimental, theoretical and production efforts in Czechoslovakia. As for the practical (production) efforts, it was of great help that the author was the leader of the team designing some of the machines and the whole line for green crop fractionation, and took part in testing the line and putting it in operation. The Škoda Company (at its Ejpovice plant) manufactured and sold seven such machine lines for GCF. Three of them were exported: one to the GDR and two to Yugoslavia. Parts of another line went to the USSR.

Description of the procedure for separating part of the protein from green crops in farm conditions using GCF

The principle of GCF is the separation of part of the juice from the plant's tissues. Mechanical separation of the juice from green forage can be divided into two stages.

1. Disintegration: breaking the integrity of the cell walls;
2. Separation:　　separation of the juice from the fibre.

These procedures can be implemented either separately (using different machines) or in a single machine. It is more advantageous to separate the two basic stages: the main reasons are that such an approach saves power and the performance is greater. The plant material can be disintegrated by means of a crusher (such as an impact grinder/hammer mill). The disintegrated material then does not need much power (about 15 kJ/kg) for separation of part of the juice by pressing at a comparatively low pressure (about 1 MPa). Special low-pressure auger presses can be used for the extraction of juice from disintegrated forage. Sugar beet slices and vine presses are suitable, after some adjustment.

The juice obtained by extraction from green fodder constitutes up to 55% of the weight of the initial material. It is a very good feed, containing about 10% dry matter of which a large proportion consists of protein; for example, in lucerne the protein constitutes 35% of the dry matter of the juice. The juice obtained can be given to pigs and also to cattle and sheep, either immediately or after timely preservation; with its high contents of protein and xanthophyll it can serve as a high-quality constituent of feed mixtures for poultry and pigs. If this juice undergoes complex processing, protein concentrates can be produced and used to replace part of the protein part of feeds, such as fish meal or soybean meal. The juice may also be processed into protein extract, which can serve as a protein constituent of human food. In these procedures, physical or chemical agents are used to coagulate the protein contained in the juice and the protein is then separated from the juice. This may include, for example, short-time heating to temperatures around 60°C or around 80°C, or the use of chemical coagulants and flocculants together with various kinds of filters, separators and driers.

To obtain dried protein–vitamin concentrate for feeding purposes, the juice extracted from the forage is quickly heated by the injection of steam during flow along a specially adjusted pipeline-coagulator. The coagulate – that is, coagulated protein together with other constituents extracted from green fodder – is then separated. This is done, for example, by means of a decantation separator, by filter presses and filters, or by a flotation tank and a sieve press. The coagulate processed in this way is already partly dewatered: it is a paste containing 30–40% dry matter. This paste can be further dried to produce a dried protein concentrate with a 90% dry matter content: this dry matter includes about 50% of crude protein, 2–3% fibre, 9% fat, about 15% minerals, about 600 mg beta-carotene per 1 kg, and 1,000 mg of xanthophyll per 1 kg. The content of metabolizable energy is between 9,000 and 10,000 kJ/kg.

A large amount of brown juice, deproteinized, is left after the separation of the coagulate from the coagulated green juice. Its dry matter content is about 6%; this dry matter contains about 20% of its volume as crude protein, and about the same proportion of sugars. It is important to utilize rationally the nutrients and thermal energy contained in this juice. It can be used as an ingredient of feed diets, but a more advantageous procedure is to thicken it in evaporators to increase dry matter content to 20–30%; treated in this way, the product can be used as molasses. Waste heat, for example that from a drier, can be used with advantage for the thickening of the juice in evaporators. The simplest way of utilizing the brown juice is to apply it to crops (manuring irrigation).

However, the brown juice can also serve for the production of fodder yeasts and ethanol. It is important to make use of the heat contained in the brown juice after heat coagulation, for example for pre-warming the green juice before coagulation.

The fodder material, such as lucerne, that remains after the extraction of the juice contains about 30–40% of its volume as dry matter (depending on the extraction procedure used) and still has enough protein (about 16% of the dry matter content) to feed ruminants; fodder conditioned in this way also has the advantage that it is more digestible. The material can either be immediately fed to livestock or ensiled (with no danger of silage juice leakage); or it can be dried. The drying of extracted disintegrated fodder requires less power. The fuel oil consumption is reduced by up to 40% and the throughput (performance) of the drier increases by up to 30%.

The cake left after extraction of juice from fodder has a relative dry matter content which is twice that of the original herbage; such a reduction of herbage moisture by extracting part of the juice – that is, mechanical dehydration – requires much less power (power consumption lower by an order) than a corresponding moisture reduction by evaporation of water in the drier. This suggests that mechanical dehydration of high-moisture herbage before drying would save power. Such an approach would be advantageous, for example, in processing the herbage harvested in a long rainy period. In this actual case the mechanical dehydration of the material would save so much power that it would be economically advantageous even without using the extracted juice for feeding purposes (the juice of high-moisture herbage is very poor in dry matter). It is advisable in such cases to skip or substantially reduce the disintegration of the material before extraction.

Summing up all this, it can be said, for example, that 100 kg of lucerne with a 20% dry matter content will yield, after extraction, 45 kg of cake with about 35% dry matter content and 55 kg of green juice with 10% of dry matter. If the cake is dried it will yield about 17 kg of dried product (lucerne meal). The green juice, coagulated with injected steam and subjected to fractionation, will give 57 kg of brown deproteinized juice and 4.8 kg of partly dehydrated concentrate-paste with about a 35% dry matter content. The drying of this paste produces 1.8 kg of dried protein–vitamin concentrate at a dry matter content of 90%.

Apart from power saving, the technology described has a number of other advantages, compared with the traditional technological systems. The result of fractionation of green crops is represented by a wide spectrum of products (xanthophyll concentrate, protein concentrates and the like) which can be used optimally in the feeding of livestock – not only cattle but also pigs and poultry – and in the food industry. The herbage nutrients can be utilized for new purposes. Beet tops can similarly be used with advantage. Another

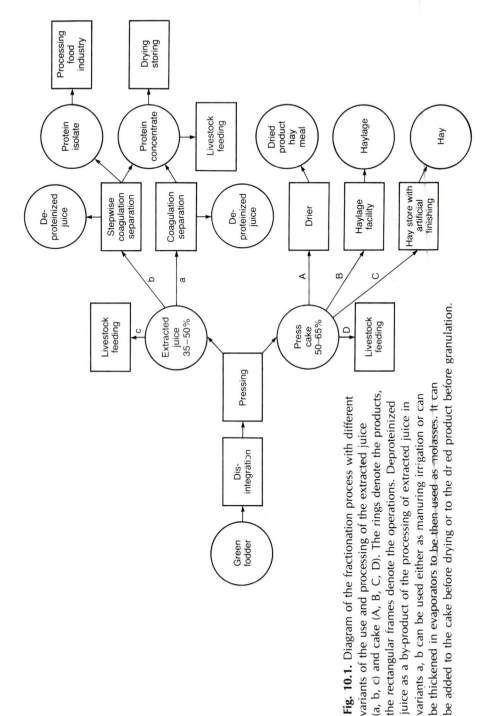

Fig. 10.1. Diagram of the fractionation process with different variants of the use and processing of the extracted juice (a, b, c) and cake (A, B, C, D). The rings denote the products, the rectangular frames denote the operations. Deproteinized juice as a by-product of the processing of extracted juice in variants a, b can be used either as manuring irrigation or can be thickened in evaporators to be then used as molasses. It can be added to the cake before drying or to the dried product before granulation.

advantage of the system is that it allows use for feeding of plants which are not normally usable for such purposes, such as potato, vines and weeds. These problems were discussed in detail in Blahovec and Řezníček, 1980, and in other publications mentioned in the References. A diagram of the GCF process is shown in Fig. 10.1.

A GCF production line for vitamin–protein concentrate

The GCF machine line is shown in Fig. 10.2. Three major types of fractionation lines can be demonstrated in the figure. The line consisting of machines

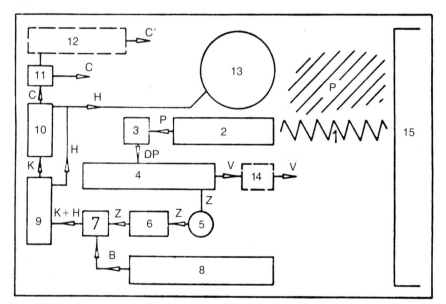

Fig. 10.2. Diagram of machine line for green crop fractionation.

P denotes the material being processed (usually chopped lucerne), (1) feed worm; (2) flinging dose conveyer; (3) crusher (hammer mill); (DP) crushed crop; (4) worm press; (Z) green juice; (5) green juice tank; (6) sieve to separate fibre; (7) coagulator; (8) steam generator; (B) steam; (K) coagulate; (H) deproteinized (brown) juice; (9) flotation tank; (10) belt press for dehydration of coagulate (a decantation separator can be used instead of the flotation tank and belt press); (11) paste granulator; (12) coagulate drier; (13) tank for deproteinized (brown) juice, possibly with a separator set for thickening the brown juice or fermentation column for the production of fodder yeasts; (V) cake; (14) cake disintegrator; (15) fodder drier or silage store or store for cake if cake is not administered immediately; (C) mechanically dehydrated vitamin–protein concentrate with dry matter contents of 30 to 35%; (C′) concentrate after hot-air drying.

1–5 represents the simplest type of fractionation line producing green juice and cake. The juice can be added to pig diets, replacing part of the protein-aceous ingredients. The cake can be fed to ruminants. The products (the juice and cake) cannot be stored without preservation. They should be administered to the livestock with the shortest possible delay. This variant of the technology is the simplest and has the most advantageous economic and power-saving parameters. However, it does not yield a wider spectrum of products that could be stored for a longer time. The line consisting of machines 1–10 produces the protein–vitamin paste (dry matter content 30–40%), cake, and brown deproteinized juice. The paste, treated with chemicals such as propionic acid, can be stored for a long time and can be used as an ingredient of monogaster diets. The line consisting of all the machines drawn in the diagram, and described in the caption to Fig. 10.2 is the most complex of all the variants. The resultant products include dried protein–vitamin concentrate and thickened brown juice (molasses). In all cases, GCF technology provides important ingredients of diets. Though obtained from the same raw material – lucerne, grass and others – these products can be included in both monogaster and ruminant diets.

Figs 10.3 to 10.9 show some of the machines in our experimental GCF line (Řezníček, 1988b).

Fig. 10.3. Fodder crusher and one-worm press with a tank for extracted juice.

Fig. 10.4. One-worm press and cake crusher.

Fig. 10.5. Steam coagulator.

Fig. 10.6. Belt press for coagulate.

Fig. 10.7. Coagulate drier.

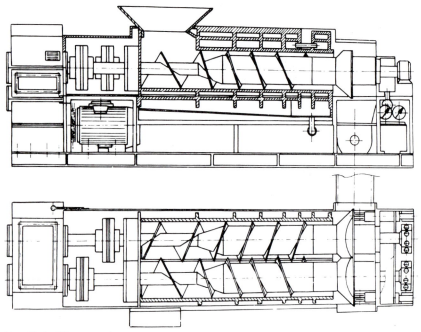

Fig. 10.8. Diagram of a two-worm press.

Fig. 10.9. A two-worm press of length 5 m, width 1.3 m, height 1.5 m; throughput performance up to 12 tonnes per hour. The power of the driving electric motor is 55 kW. (The lid of the pressure chamber has been removed.)

Physical aspects of the main part of the GCF process

Disintegration (crushing) of the plants to be processed is the first step in GCF. Its main purpose is to break the integrity of the cell walls and to release the protoplasm into the intercellular space and into the space between plant tissue fractions. Physical analysis of the disintegration of plant tissues shows that the integrity of cell walls can be best broken in those tissues where a pressure gradient has been formed. This can be achieved in several ways – for example, when compressing the plant tissues by means of tools with an irregular surface, by impact and by shear stress, the non-homogeneity of the plant material helping to form non-homogeneous pressure fields. Dynamic action on the material is more efficient in this process than static action, though the stress values are the same. The physical principles of the crushers used are associated with this. The crushers either compress the plants between rollers or extrude them through openings, or act on them by frictional force. There are also various types of impact grinders and choppers. Fomin (1978) calculated the theoretical value of the energy needed to mechanically break the cell walls in 1 kg of compressed lucerne. The result of his calculation is that this energy is 0.17 kJ/kg. The specific energy consumed for disintegration of herbage in laboratory trials is about 5 times greater, and under farm conditions it is substantially higher still. For instance, for the impact grinder disintegrator this value is 15–20 kJ/kg, and in very intensive disintegration to produce pulp, it is 50–150 kJ/kg.

The following physical characteristics are involved in the conditions under which the cell walls are broken, as follows from theoretical analysis:

E Young's modulus of cell wall;
ν Poisson's ratio of cell wall;
σ_p tensile strength of cell wall;
h cell wall thickness;
a thickness of one layer of cells;
b cell width;
n number of cells in half the sample width;
$2L$ cell layer width;
i cell rank, counted from the middle of the layer.

Fomin (1978) derived the following relations for pressure p_i in the ith cell:

$$p_i = \frac{\pi^6 (3 - \nu)}{512(1 - \nu)} E \left(\frac{h}{a}\right)^4 \left(\frac{\xi}{n}\right)^3 [n^2(n + 1)^2 - i^2(i + 1)^2] -$$

$$\left[\left(\frac{\pi^2 a^2}{(1 - \nu) h b}\right) \left(\frac{\xi}{n}\right) - \frac{\pi^6}{12(1 - \nu^2)}\right]\left(\frac{\xi}{n}\right)\left[\frac{(n - i)(n + 1 - i)}{8} E\left(\frac{h}{a}\right)^4\right] \quad [10.1]$$

where

$$\xi = \frac{32\nu + \sqrt{(32\nu)^2 + 32(1+\nu)(2-\nu)(1-\nu^2)\pi^6 \dfrac{L^2}{a^2} \dfrac{\sigma_p}{E}}}{2(1+\nu)(2-\nu)\,\pi^4 \dfrac{Lh}{2}} \qquad [10.2]$$

and for the deformation work needed to disintegrate a unit volume of herbage by compression:

$$W_{dt} = \left(\frac{(3-\nu)(1-\nu)}{2(2-\nu)^2}\right)\left(\frac{h}{b}\right)\left(\frac{\sigma_p^2}{E}\right) \qquad [10.3]$$

Calculations similar to those for a cell layer compressed between two plates were also performed by Fomin (1978) for a cell layer deformed by shear between two plates. The shear plane is formed by the planes of the compressing plates; the direction of the shear deformation is perpendicular to the length of the layer. The following expression was found for the critical shear stress in layer τ_k at which the side walls of the cells fail:

$$\tau_k = \left[\tau_{lk} + \frac{3\sqrt{2}}{\psi_1(\nu)}(6+4\nu-\nu^2)\sigma_p\right]\frac{h}{a} + \frac{E}{1-\nu^2}\left[(tg\theta - \sin\theta)\frac{h}{b}\right] \qquad [10.4]$$

where

$$\tau_{lk} = \frac{\sqrt{2}\pi^2\,E\,h^2}{3a^2(1-\nu^2)} \qquad [10.5]$$

$$\psi_1(\nu) = \frac{1}{2}\left[(31+18\nu-13\nu^2) + \right.$$

$$\left. \sqrt{3200 + 1914\nu + 383\nu^2 - 450\nu^3 + 255\nu^4}\right] \qquad [10.6]$$

$$\theta = \frac{2(1+\nu)}{E\psi_1(\nu)}\{\tau_{lk}\psi_1(\nu) + [1-\nu+3\sqrt{2}(6+4\nu-\nu^2)\sigma_p]\} \qquad [10.7]$$

For the specific deformation work (related to unit volume) the expression is:

$$W_{ds} = \frac{\sigma_p^2}{4E}\left[\left(1+\frac{1}{m_y}\right)\frac{h}{a}\psi_4(\nu) + \frac{1}{2\psi_3^4(\nu)}\left(1+\frac{1}{m_x}\right)\frac{h}{b}\frac{\sigma_p^2}{E^2}\right] \qquad [10.8]$$

where m_x is the number of cells in the layer in the direction of the shear and m_y is the number of cells in the layer in direction perpendicular to the direction of the shear. The functions $\psi_3(\nu)$ and $\psi_4(\nu)$ are given by the relations:

$$\psi_3(\nu) = \psi_1(\nu)\left\{2(1 + \nu)[1 - \nu + 3\sqrt{2}\,(6 + 4\nu - \nu^2)]\right\}^{-1} \qquad [10.9]$$

$$\psi_4(\nu) = \frac{3029 + 6328\nu + 3130\nu^2 - 560\nu^3 - 427\nu^4 + 64\nu^5}{\psi_1^2(\nu)} \qquad [10.10]$$

The basic operation in GCF is the extraction of part of the liquid fraction from the fodder. The following main physical quantities are involved in the theoretical description of this procedure:

ρ_o volume weight of herbage, for example chopped lucerne, at the beginning of extraction;

ρ_h volume weight of herbage in the state of hydromass; that is, after disintegration;

$\overline{l}$ characteristic length of chopped particles;

ρ_o f $(\overline{l})$ – dependence of volume weight on the characteristic length of chopped particles;

γ volume compressibility of the herbage (chopped material); $\gamma = -dV/Vdp$, where V is volume, dV is volume change caused by change in pressure dp;

p   pressure in the pressure chamber;

λ_j degree of release of the juice, which is the quotient of the weight of extracted juice and the weight of material (chopped lucerne) from which the juice was extracted;

W_i specific energy for the extraction of the juice; it is the energy needed to release a unit weight of juice by extraction;

Δp pressure gradient in the layer being compressed, needed for the flow of the juice; $\Delta p = dp/dr$;

K   coefficient of filtration of juice through the layer of herbage being compressed $K = -v_r/\Delta p$;

v_r flow rate in the direction of r;

f_1 coefficient of friction of the material against the worm of the press;

f_2 coefficient of friction of the material against the pressure chamber of the worm (this chamber consists of a screen; the juice is pressed through the meshes).

During the process of extraction in a single-worm press, the volume of material contained in one turn of the worm will change by ΔV in time Δt. ΔV is the sum of the volume ΔV_1 of extracted juice (liquid fraction) and the change in volume ΔV_2 of the pressed fodder, as produced by compression in the press.

$$\Delta V = \Delta V_1 + \Delta V_2 \qquad [10.11]$$

In the case of a single-worm press:

$$\Delta V_1 = \pi D_o \, v_r \, s \, \Delta t \qquad\qquad [10.12]$$

where D_o is the outer diameter of the helix;
       v_r is the flow rate of the liquid fraction (juice) in the radial direction; and
       s is the lead of the helix: $s = \pi \, D_o \, tg\alpha$

According to Darcy's law:

$$v_r = -K \, dp_r/dr \qquad\qquad [10.13]$$

where dp_r/dr is the pressure gradient Δp_r in radial direction. If D_k denotes the inner diameter of the worm helix on whose surface there is the pressure p_k, then it holds, with simplification, that:

$$\Delta p_r = p_k/(D_o - D_k) \qquad\qquad [10.14]$$

and:

$$\Delta V_1 = -2\pi k \, p_k \, s\Delta t \, D_o/(D_o - D_k) \qquad\qquad [10.15]$$

For the shift Δx of the material in time Δt inside the press in the direction of the axis of the worm helix it holds that $\Delta x = s \, \Delta t \, n$, where n is the frequency of turning of the worm helix. Then:

$$\Delta V_1 = -2\pi K \, p_k \, \Delta x \, D_o \, \phi/n \, (D_o - D_k) \qquad\qquad [10.16]$$

The coefficient ϕ expresses the slipping of the material in tangential direction along the walls of the pressure chamber.

On the basis of the definition of volume compressibility,

$$\Delta V_2 = -\gamma V \, (\partial p/\partial x)\Delta x \qquad\qquad [10.17]$$

where V is the volume being compressed, p is the pressure in the pressure chamber, and $(\partial p/\partial x) \, \Delta x$ represents the pressure increase generated during the operation of the press between two points Δx apart.

The change in the volume of the material being pressed is given by the dimensional characteristics of the press parts in which the compression occurs. Depending on the design of the press, the following have a bearing on this change ΔV_a: the change in the diameter of pressure chamber ΔV_{a1}, the change in the diameter of the worm hub ΔV_{a2}, and the change in the lead of the helix of the worm ΔV_{a3}. The contribution of the change in volume of the compressed material ΔV_{a4}, produced by the change in slipping should also be considered. If the press is to work in an equilibrium, $\Delta V = \Delta Va$ must hold along the whole pressure chamber.

These data, together with knowledge concerning the pressure field in the pressure chamber, can be used for estimation of the specific energy whose value is of the order of the kJ/kg units. However, the mechanical efficiency of the drive and run of the press is poor and in the worm presses actually used in practice, W_i is between 10 and 20 kJ/kg.

The coagulation of the proteinaceous components of the extracted plant juice is another important operation in GCF. In coagulation by warming of the juice, the purpose is to obtain particles of coagulated protein of sufficient strength with a surface energy permitting the formation of larger cohesive agglomerates of these particles as a result of coalescence. This requirement ensues from the need to separate these agglomerates from the protein-free juice (for example, by flotation) and then to dehydrate them mechanically (for example, by means of a band screen filter press). If the structure of the agglomerates is not strong enough the screens become clogged and the mechanical dehydration of the coagulate by means of the filter press is not effective.

The possibility of separating the largest possible amount of protein from the lucerne juice by heat coagulation and to obtain a coagulate of the required properties, taking into account the need to save power, depends mainly on coagulation temperature, time of heating, pH of the juice, age of the juice and quality of the lucerne crop used. From these points of view, the best pH is 9.4, the best temperature is 80°C and the best heating time is several seconds. The juice should be coagulated as soon as it is extracted. The lucerne crop must be fresh and free of any contaminants. To obtain the required pH of 9.4, ammonia liquor is added to the juice. Hence, a through-flow coagulator in which steam is injected into the flowing juice is suitable for rapid warming and processing of the juice when there is no long delay after extraction.

The theoretical model derived from the theory of statistical physics (Landau and Lifshits, 1964; Levič 1954, Řezníček and Lejčková, 1988) can be used for physical analysis of coagulation, because the coagulated particles are regarded as a statistical system of particles in a thermodynamic equilibrium at a given temperature. The distribution of coagulated particles in dependence on their radius can be expressed as:

$$w = \frac{1}{kT} \, e^{-(E/kT)} \, \frac{dE}{dr} \qquad [10.18]$$

In this expression, w is the relative frequency of particles with diameter r at temperature T. The parameter E is the potential energy of the particles and k is Boltzmann's constant. For E it is assumed that:

$$E = K \, v \qquad [10.19]$$

where K is the bond energy per unit volume, holding the particles together, and v is the particle volume.

In the approximation of particle form, which is generally irregular, the particle is assumed to be a sphere.

$$E = \frac{4}{3}\pi r^3\, K \qquad\qquad [10.20]$$

By substitution for E in the relation for w, we obtain the relative number frequency of the spherical particles with radius r at temperature T:

$$w = \frac{4\pi K}{k\,T}\, r^2 e^{-\frac{4\pi}{3} r^3 K/kT} \qquad\qquad [10.21]$$

and the number of particles N_h in the interval of $[r - h, r]$

$$N_h = N \int_{r-h}^{r} w\, dr \qquad\qquad [10.22]$$

where N is the total number of particles in the system's volume.

$$N_h = N \left[e^{-\frac{4\pi}{3}(r-h)^3 K/kT} - e^{-\frac{4\pi}{3} r^3\, K/kT} \right] \qquad\qquad [10.23]$$

$$\bar{r} = \int_0^\infty r\, w\, dr = \frac{4\pi K}{k\,T} \int_0^\infty r^3\, e^{-\frac{4\pi}{3}\frac{r^3 K}{k\,T}} dr = \int_0^\infty e^{-\frac{4\pi r^3 K}{3kT}} dr$$

$$= \sqrt[3]{\frac{3\,kT}{4\pi K}}\, G \qquad \text{(G is a constant)} \qquad\qquad [10.24]$$

When studying the rise of the critical nucleus of a coagulated particle it is also necessary to take into account the surface energy of nucleus σ. Hence, the total potential energy R is:

$$R = -E + 4\pi\, r^2 \sigma \qquad\qquad [10.25]$$

From the theory of fluctuations (Landau and Lifshits, 1964) it follows that the distribution function of the nucleus in dependence on radius f(r) obeys:

$$f(r) \sim e^{-\frac{R}{kT}} \qquad\qquad [10.26]$$

Function f(r) is the number of nuclei of spherical shape and radius r in the whole volume of the system. The size of the critical nucleus corresponds to the maximum of R.

$$r_k = 2\,\frac{\sigma}{K} \qquad\qquad [10.27]$$

The kinetic equation for the time distribution function f(r,t) of the number of the generated nuclei – the so-called Fokker–Planck equation (Lifshits and Pitaevsky, 1979) – is used, reading:

$$\frac{\partial f(r,t)}{\partial t} = -\frac{\partial S}{\partial r} \qquad [10.28]$$

where

$$S = \frac{dr}{dt} f(r,t) - B \frac{\partial f(r,t)}{\partial r} \qquad [10.29]$$

is the flow density in the 'nucleus radius space'. Coefficient B is determined by the condition of thermodynamic equilibrium, when f(r,t) = f(r) and S = 0. Then:

$$B = -kT \frac{(dr/dt)}{(dR/dr)} \qquad [10.30]$$

The steady-state solution to the equation meets the condition S = constant. This constant value yields the number of nuclei in $1 \, cm^3$ per s whose radius has reached the critical level. The steady-state solution to the equation (Lifshits and Pitaevsky, 1979) gives:

$$S = 2 \sqrt{\frac{\sigma}{kT}} \, B(r_k) \, f \, (r_k) \qquad [10.31]$$

The magnitude of $B(r_k)$ can be determined by applying the procedure described by Lifshits and Pitaevsky (1979) for crystallization from a saturated solution to the case of clustering of coagulated particles. Let c_∞ be the concentration of coagulate far away from the nucleus. Further growth of the nucleus whose radius r has exceeded the critical value – that is, $r > r_k$ – takes place by diffusion of the coagulated particles to the nucleus. The distribution of the concentration in the neighbourhood of the nucleus $c(\mathcal{L})$ where $\mathcal{L}$ is the distance from the centre of the nucleus, radius r, is given by the equation

$$\frac{\partial c(\mathcal{L})}{\partial t} = -D \frac{1}{\mathcal{L}} \frac{\partial^2}{\partial \mathcal{L}^2} \, \mathcal{L}c(\mathcal{L}) \qquad [10.32]$$

where D is the diffusion coefficient.

The steady-state solution to $\frac{\partial c}{\partial t} = 0$ is

$$c(\mathcal{L}) = c_\infty - (c_\infty - c_r) \frac{r}{\mathcal{L}} \qquad [10.33]$$

where c_r is concentration at the surface of a nucleus with radius $\mathcal{L} = r$.

The diffusion flow I towards the nucleus is given by the relation:

$$I = 4\pi r^2\, D\frac{dc}{d\mathcal{L}} = 4\pi D\, r\, (c_\infty - c_r) \qquad [10.34]$$

If c_0 is the mean concentration of saturated solution, then according to Landau and Lifshits (1964) the critical radius of the nucleus when it is separated from the saturated solution is:

$$r_k = \frac{2\sigma v_m\, c_0}{kT\, (c_\infty - c_0)} \qquad [10.35]$$

where v_m is the molecular volume of the coagulated protein.

Finally,

$$I = 4\pi D\, (c_\infty - c_0)\, (r - r_k) \qquad [10.36]$$

Since concentration c is the number of coagulated protein molecules in a unit volume, I is the number of molecules which become attached to a surface of $4\pi r^2$ of the supercritical nucleus per second. Consequently, the rate at which the radius r of the supercritical nucleus changes (dr/dt) will be:

$$\frac{dr}{dt} = \frac{I\, v_m}{4\pi r^2} = \frac{D\, v_m\, (c_\infty - c_0)\, (r - r_k)}{r^2} \qquad [10.37]$$

Using the relation for R will yield

$$\frac{dR}{dr} = 8\,\pi\sigma\, r - 4\pi r^2 K = 4\pi Kr(r_k - r) \qquad [10.38]$$

By substituting into the equation for B, we arrive at:

$$B(r_k) = \frac{D v_m^2 c_0}{4\pi r_k^3} \qquad [10.39]$$

The number of nuclei S which have reached critical size per cm^3 and per s is then:

$$S \sim 2\sqrt{\frac{\sigma}{kT}}\,\frac{D v_m^2 c_0 K^3}{32\pi\sigma^3}\,\exp\left[-\frac{16\pi\sigma^3}{3kT\, K^2}\right] \qquad [10.40]$$

This theory holds goods on the assumption that the overall volume of all nuclei is so small that the rise of any nucleus owing to fluctuations and further growth take place independently of other nuclei. In the later stage when the oversaturation of the solution is small — that is, $c_o \rightarrow c_\infty$ — the nature of the process changes. Fluctuations produce practically no further nuclei, and further growth of the radius of the cluster r is given by coalescence — that is, by smaller clusters being absorbed by the larger ones. Further growth of cluster radius is then controlled only by the diffusion of the smaller clusters to the larger ones. From this point of view, the process of coalescence is somewhat similar to the growth of the critical nuclei, the difference being that, in the case of coalescence, the clusters already exist and their generation due to fluctuation does not have to be considered. The growth process then comes to a stop if the saturation of the solution with small clusters decreases. The condition of thermodynamic equilibrium can then be assumed to exist with the mean equilibrium radius r. The radius of the clusters is now considered to be sufficiently large to allow neglecting, for the sake of simplicity, $4\pi r^2 \sigma$ relative to E. Rate S also decreases with increasing temperature; the process of coagulation is terminated and the thermal oscillations disintegrate the generated clusters.

References

Blahovec, J. and Řezníček, R. (1980) *Frakcionace píce* (Fodder fractionation). VŠZ-MON, Prague.

Fomin, V. J. (1978) Vlazhnoe fraktsirovanie zelenykh kormov (Wet fractionation of green fodder). Izdatel'stvo Rostovskogo Universiteta.

Landau, L. D. and Lifshits, E. M. (1964) Statisticheskaya fizika (Statistical physics). Nauka, Moscow.

Lévič, V. G. (1954) Úvod do statistické fyziky (Introduction to statistical physics). Academia, Prague.

Lifshits, E. M. and Pitaevsky, L. T. (1979) Fizicheskaya kinetika (Physical kinetics). Nauka, Moscow.

Pirie, N. W. (1978) *Leaf Protein and Other Aspects of Fodder Fractionation*. Cambridge University Press, Cambridge.

Řezníček, R. (ed.) (1988a) *Physical Properties of Agricultural Materials and Products*. Hemisphere Publishing Corporation, New York.

Řezníček, R. (1988b) Research into Green Crop Fractionation Conducted at the Faculty of Mechanization of the University of Agriculture, Prague. In: Řezníček, R. (ed.), *Physical Properties of Agricultural Materials and Products*, Hemisphere Publishing Corporation, New York, 1139–44.

Řezníček, R. and Lejčková, K. (1988) On the problem of studying the coagulation of green plant juice. In: Řezníček, R. (ed.), *Physical Properties of Agricultural Materials and Products*. Hemisphere Publishing Corporation, New York, 196–200.

Chapter 11

Principles of Abattoir Design to Improve Animal Welfare

Temple Grandin

Introduction

Restrainers, races, holding pens and unloading ramps in abattoirs must be properly designed to facilitate animal movement, prevent bruises and minimize stress. The livestock industry loses millions of dollars annually due to bruises and death losses (Livestock Conservation Institute, 1988; Marshall, 1977). Stress-related meat quality problems such as dark cutting (DFD) and pale soft exudative (PSE) meat also cause huge losses (Canadian Meat Council, 1980). Well-designed abattoir facilities will also greatly improve animal welfare. Engineers must always remember that good facilities provide the tools to make humane handling possible, but management is essential to enforce good handling practices. Engineering is not a substitute for management.

Different countries will have specific requirements for facilities; for example, truck size will affect the size of the holding pens. Space and facilities must also be designed for specialized functions, such as weighing, animal identification, and washing. The engineer must be familiar with the specific requirements of the country in which the facility is being constructed.

Livestock behaviour facility design

Vision

Livestock have wide-angle vision. The visual field of cattle and pigs is over 300° (Prince, 1977). In sheep the visual field varies from 191–306°, depending on fleece thickness on the head. Races, crowd pens, and unloading ramps should have solid fences to prevent animals from seeing distractions outside the fence with their wide-angle vision (Grandin, 1980a, 1982; Rider *et al.*, 1974). Moving objects outside the fence will frighten livestock. The use of solid fences in races and crowd pens is especially important if animals are not completely tame, but solid fences in races and crowd pens will also help keep tame animals calmer.

Animals will often balk at a sudden change in fence construction or floor texture (Lynch and Alexander, 1973). Puddles, shadows, drains and bright spots will also impede animal movement. Poor depth perception may explain why livestock balk at many things. Livestock can perceive depth when they are still and have their heads down (Lemman and Patterson, 1964), but their ability to perceive depth while they are moving with their heads up may be poor (Hutson, 1985). To see depth accurately the animal has to stop and put its head down.

Indoor handling facilities should have even, diffuse lighting that minimizes shadows. Cattle, pigs and sheep have a tendency to move more easily from a dimly illuminated area to a more brightly illuminated area (Grandin, 1982; Hitchcock and Hutson, 1979; Kilgour, 1971; VanPutten and Elshof, 1978). At night or in enclosed facilities, lamps can be used to attract animals into races (Grandin, 1982). The lights should illuminate the floor and must not shine into the eyes of approaching animals. Livestock are more likely to balk if they are forced to move towards blinding sunlight. Pigs reared indoors under artificial illumination preferred to walk up a ramp illuminated at 80 lux (Phillips *et al.*, 1987); this was similar to the illumination of their living quarters. A dimly illuminated ramp with less than 5 lux was avoided and there was also a tendency to avoid an excessively bright ramp illuminated at 1200 lux (Phillips *et al.*, 1987).

Noise

Livestock have sensitive hearing and they are stressed by excessive noise (Kilgour and deLangen, 1970; Kilgour, 1983). They are specially sensitive to high-frequency sound around 7000–8000 Hz (Ames, 1974). Humans are most sensitive at 1000–3000 Hz (Ames, 1974). In steel facilities, gate strike posts should have rubber stops to reduce noise and in the stunning area, the shackle return should be designed to prevent clanging and banging. Air exhausts on pneumatically powered gates should be piped outside (Grandin, 1983b). If hydraulics are used to power gates or conveyors, the motor and pump should be located away from the animals and all hydraulic motors and plumbing should be designed to minimize noise; high-pitched sound from a hydraulic system is very disturbing to cattle. Cattle held overnight in a noisy yard close to the unloading ramp were more active and showed more bruising compared to cattle in a quieter pen (Eldridge, 1988).

Experience and genetic effects

An animal's previous experiences at the farm or ranch of origin will affect its behaviour at the abattoir. Cattle which have been handled roughly at the feedlot of origin become more agitated and are more difficult to handle than cattle which have received gentle treatment.

At one abattoir, playing music throughout the stockyard and race areas reduced excitement and agitation. The cattle became accustomed to the music in the holding pens and it provided a familiar sound which helped mask the sounds of abattoir equipment.

Pigs reared in an extremely barren environment with a lack of stimulation will be more excitable than pigs reared with additional stimulation, such as rubber hose toys or people walking in the pens (Grandin, 1989a). Pigs reared in confinement with a lack of stimulation were more difficult to load than pigs reared outside (Warriss *et al.*, 1983). Playing a radio in the fattening pens will reduce the animals' reactions to sudden noises such as a door slamming, and possibly stress could be reduced if animals were exposed to abattoir sounds on the farm – animals will readily adapt to reasonable sound levels. Ames (1974) found that continuous exposure to either instrumental music or miscellaneous sounds during fattening, improved weight gain of sheep at 75 dB and reduced weight gain when played at 100 dB. Exposure to reasonable sound levels during fattening will help reduce fear reactions to sudden unexpected noises.

Genetics is also an important determinant of how animals will behave at the abattoir. Brahman and Brahman-cross cattle are more excitable than Hereford or Angus (Tulloh, 1961). Since the mid-1980s the author has observed increasing problems with very excitable nervous pigs which are very difficult to handle at the abattoir. They have extreme shelter-seeking behaviour (flocking together), and the animals will not separate from the group and move up the race (Grandin, 1989b). This problem has increased and is often most evident in hybrid pigs which are selected for leanness and high productivity. Some of these animals are so 'crazy' that it is almost impossible to handle them humanely in a conventional race and crowd pen system. When these pigs become excited, handlers may over-use electric prods and meat quality will be reduced. Designing special facilities to handle 'crazy' pigs will be very expensive. This problem should be corrected at its source by selective breeding for lean pigs with a calm temperament. Engineers should avoid the temptation to try to treat the symptoms of a problem with engineering, rather than correcting the problem at its source.

Layout of lairages and stockyards

Long, narrow pens are recommended in stockyards and lairages in abattoirs
(Grandin, 1980a, b; Kilgour, 1971). One advantage of long, narrow pens is
more efficient animal movement: animals enter through one end and leave
through the other. To eliminate 90° corners, the pens can be constructed
on a 60–80° angle (Figs 11.1, 11.2, 11.3). Each pen gate should be longer
than the width of the alley, so that it opens on an angle to eliminate the sharp
corner (Fig. 11.1).

Long, narrow pens maximize lineal fence length in relation to floor area
and this may help reduce stress (Kilgour, 1978; Grandin, 1980a, b). Cattle
and pigs prefer to lie along the fenceline (Grandin, 1980b; Stricklin *et al.*,
1979). Observations indicate that long, narrow pens may help reduce fighting
(Kilgour, 1976). Government regulations in some countries may require walkways in between the pens for observation of animals prior to slaughter. The
layout remains the same except a 1 m wide walkway is placed between every
other pen.

The size of the holding pens required for an abattoir is going to be at
least partially dictated by size of the trucks. When small groups of animals
are handled, block gates can be used in a long, narrow pen to keep different
groups separated. Minimum space requirements for holding fattened, feedlot
steers for less than 24 hours are $1.6\,m^2$ for hornless cattle and $1.85\,m^2$ for
horned cattle (Grandin, 1979; Midwest Plan Service, 1980), and $0.5\,m^2$ for
slaughter-weight pigs and lambs. During warm weather pigs require more
space. Wild, extensively raised cattle may also require additional space.
However, providing too much space may increase stress because wild cattle
tend to pace in a large pen. Enough space must be provided to allow all
animals to lie down at the same time. In countries with large trucks, larger
pens and wider alleys will be required. To avoid bunching and trampling 25 m
is the maximum recommended length of each holding pen, unless block gates
are installed to keep groups separated, shorter pens are usually recommended.

Recommended alley and pen width will vary depending on the number
of animals that will be moving through the facility. In pig and sheep abattoirs
where the line speed is under 400 animals per hour or where trucks holding
less than 80 animals are used, pen width should be 2 m and the alley width
should be 1.3 m. In abattoirs served by trucks which hold over 150 pigs or
sheep per load, pen width should be increased to 3.0–4.2 m and alley width
will vary from 2.5–3 m. In large cattle abattoirs slaughtering 60 or more cattle
per hour, 3 m alleys are recommended. Smaller abattoirs can reduce alley
width to save space.

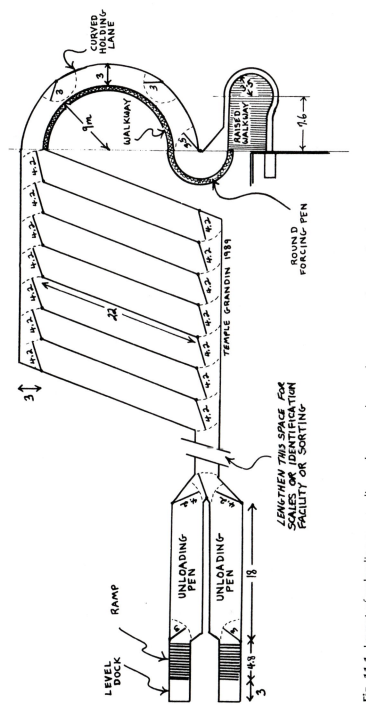

Fig. 11.1. Layout of unloading ramps, diagonal pens and curved race system for a cattle abattoir.

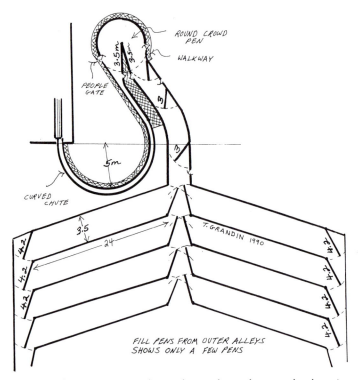

Fig. 11.2. Curved race system and round crowd pen for a cattle abattoir.

Fig. 11.3. Gates on diagonal pens should be longer than the width of the alley to eliminate a 90° corner.

Avoid mixing strange animals

To reduce stress, prevent fighting, and preserve meat quality, strange animals should not be mixed shortly before slaughter (Barton-Gade; 1985; Grandin, 1983a; Tennessen and Price, 1980). Solid pen walls between holding pens prevent fighting through the fences. Solid fences are especially important in lairage and stockyard pens if wildlife such as deer, elk, or buffalo are handled.

To keep pigs separated presents some practical problems. In the USA, pigs are transported in trucks with a capacity of over 200 animals; however, they are fattened in much smaller groups. Observations at UK abattoirs indicate that mixing 200 pigs from three or four farms resulted in less fighting than mixing 6–40 pigs. One advantage of the larger group is that an attacked pig has an opportunity to escape. Price and Tennessen (1981) found a tendency towards more DFD carcasses and hence more stress when small groups of 7 bulls were mixed compared to larger groups of 21 bulls.

In Denmark, the design of the pig lairage at the abattoir is very specialized (Fig. 11.4). Pigs are held in long, narrow pens equipped with manual push gates and a powered push gate moves pigs up the alley to the stunner. This system was invented by T. Wichmann of the Danish Meat Research Institute. The Danes have also developed automated block gates within the long, 2 m wide pens to keep small groups of 15 pigs in separate groups (Barton-Gade, 1989). A powered gate moves along each pen and brings groups of 15 pigs up to the main drive alley. After a group of pigs is brought to the alley the powered push gate is reversed and moved back down the holding pen to bring up the next group of 15 pigs. It is raised so that it can pass over the animals and then lowered to bring them out of the pen. Block gates spaced at intervals along each pen keep three or four groups of 15 pigs separated. Since the biparting block gates swing back into the animals, overloading of each 15-head compartment is prevented (if the compartments are overloaded, the system will not work). This system can deliver a steady flow of 400–500 pigs per hour to the main drive alley and it greatly reduced damage and bruises caused by fighting. Each group of 15 pigs were pen mates on the farm. A disadvantage of this system is very high cost, and it may not work with some of the new genetic lines of pigs which are highly excitable and difficult to handle. The Danish abattoir that had this system had calm placid pigs which moved easily.

When strange bulls are mixed, physical activity during fighting increases DFD meat. The installation of either steel bars or an electric grid over the holding pens at the abattoir, prevented dark cutting in bulls (Kenny and Tarrant, 1987). These devices prevent mounting. The electric grid should only be used with animals that have been fattened in pens equipped with an electric grid. In Sweden and other countries where small numbers of bulls are fattened, individual pens are recommended at the abattoir (Puolanne and

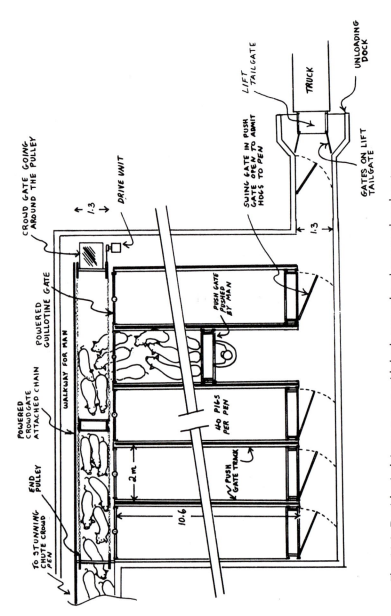

Fig. 11.4. Danish pig lairage system with both manual and powered push gates.

Aalto, 1981). In some European abattoirs, the holding area consists of a series of single file races which lead to the stunner. Bulls are unloaded directly into the races and each bull is kept separated by guillotine gates.

Flooring and fence design to reduce injuries

Floors must have a non-slip surface (Grandin, 1983a; Stevens and Lyons, 1977). For cattle, concrete floors should have deep 2.5 cm V-grooves in a 20 cm square or diamond pattern. Do not use the deep groove pattern for cattle living quarters such as cubicle housing or tied housing for milking cows. Prolonged standing for several weeks on the rough floor will damage the animals' feet. For pig and sheep abattoirs imprint the wet concrete with a stamp constructed from expanded steel mesh which has a 3.8 cm long opening (Grandin, 1982).

Concrete slats may be used in holding pens, but the drive alleys should have a solid concrete floor. Precast slats for cattle or swine confinement buildings will work well, and the slats should be specially ordered with a grooved surface. Slats or gratings used in pig and sheep facilities should face in the proper direction as sheep move more easily when they walk across the slats instead of parallel with them (Hutson, 1981; Kilgour, 1971), and the floor appears more solid when the animals walk across the slats. To facilitate animal movement, the animals must not be able to see light or reflection off water under the slats.

Animals will balk at sudden changes in floor texture or colour, so all flooring surfaces should be uniform in appearance and free from puddles (Lynch and Alexander, 1973). In facilities that are washed, there should be concrete kerbs installed between the pens to prevent water in one pen from flowing into another. Drains should be located outside the areas where animals walk as livestock will balk at drains or metal plates across an alley (Grandin, 1987). Flooring should not move or jiggle when animals walk on it. Flooring that moves causes swine to balk (Kilgour, 1988).

Cattle and sheep can have bruises on the inside even though the hide is undamaged. Pigs are slightly less susceptible to bruising, but their meat quality is decreased when they become excited or hot. Edges with a small diameter will cause severe bruises. Never use steel angles or I-beams to construct livestock facilities: animals bumping the edges will bruise. Round pipe posts and fence rails are recommended. Surfaces which come into contact with animals should be smooth and rounded (Grandin, 1980c; Stevens and Lyons, 1977) and any exposed sharp ends of pipes should be bevelled or covered to prevent gouging. Areas which have completely solid fences should have all posts and structural parts on the outside away from the animals. An animal rubbing against a smooth flat metal surface will not bruise. A practical way to determine if a fence will cause bruising is to rub your shoulder against it; if it snags your shirt or jacket, it will cause a bruise.

All gates should be equipped with tie-backs to prevent them from swinging out into the alley and guillotine gates should be counterweighted and padded on the bottom with conveyor belting or large diameter hose (Grandin, 1983a).

Design of races, crowd pens and unloading ramps

Races

All races should have solid outer fences, and a curved single file race is especially recommended for moving cattle (Grandin, 1980a; Rider *et al.*, 1974). An inside radius of 5 m is ideal for cattle: if a shorter radius is used there must be a section of straight race at the junction between the single file race and the crowd pen. If the race is bent too sharply at the junction between the single file race and the crowd pen, it will cause animals to balk (Grandin, 1987). The race entrance must not appear to be a dead end.

Figures 11.5 and 11.6 illustrate cattle races which are built correctly: walkways for the handler should run alongside the race; the walking platform

Fig. 11.5. Cattle readily walk up a curved race with solid sides.

Fig. 11.6. The handler walkway should run along the inside radius.

should be 1 m from the top of the fence; the use of overhead walkways should be avoided. In abattoirs with restricted space a serpentine race system (Grandin, 1984) can be used, and if smooth continuous bends are used, the radius can be reduced to 1.5 m (Fig. 11.7). The system must be laid out so that the animals standing at the race entrance can see a minimum of three body lengths up the race.

Curved races provide little or no advantages for pigs, because they have a strong urge to move forward in a race. A straight race will work efficiently. A race system at an abattoir must be long enough to ensure a continuous flow of animals to the stunner, but not be so long that animals become stressed waiting in line.

A common mistake is to build races too wide. Table 11.1 shows the correct lengths for cattle, pig, and sheep races. Cattle and sheep races should have straight sides. Jamming, and pigs rooting under each other, can be reduced by narrowing the bottom of the race.

In pig abattoirs, two races are sometimes built side by side, because the animals will enter more easily (Grandin, 1982). The outer walls of the race are solid, but the inner fence in between the two races is constructed from bars. This enables the pigs to see each other and promotes following behaviour. However, this system still causes stress at the stunner because pigs on one side have to wait. Stress could be greatly reduced by installing two stunners. This would enable two lines of pigs to continuously move forward, and is economically viable for large plants slaughtering over 500 pigs per hour.

Table 11.1. Recommended single file race length for abattoirs.

Line speed (per hour)	Cattle (m)	Pigs and sheep (m)
20–75	14	6
80–150	24	11
150–300	30	11
300–500	n/a	11
>500	n/a	

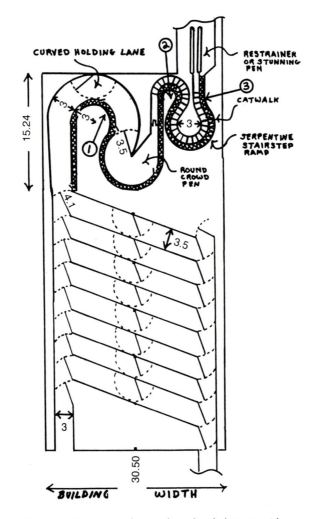

Fig. 11.7. Serpentine layout for a beef abattoir with restricted space.

A future possibility is to eliminate traditional single-file races. Figure 11.8 illustrates an idea for multiple stunners with several parallel races. Funnelling the pigs down into single file is eliminated. This system would be very expensive, but may be required if pig breeders fail to select pigs for ease of handling. Swedish and Danish engineers have proposed putting pigs in individual boxes on a conveyor. This idea is so expensive that it is not practical. In Denmark there has been experimentation with group containers which hold a pen of pigs.

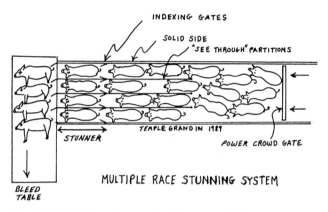

Fig. 11.8. Multiple pig races leading to multiple stunners would prevent bunching and stress.

Crowd pens

Round crowd pens are very efficient for all species. The recommended radius for round crowd pens is 3.5 m for cattle, 1.83 m to 2 m for pigs, and 2.4 m for sheep. All crowd pens must have solid fences. For smooth animal movement never construct a building wall at the junction between the single-file race and the crowd pen. The race must extend from the wall a minimum of two body lengths.

Cattle and sheep crowd pens should have a funnel design with one straight fence and the other fence on a 30° angle (Meat and Livestock Commission, no date; Figs 11.1 and 11.2). The funnel design should not be used with pigs, as they will jam at the race entrance and jamming is very stressful for pigs (VanPutten and Elshoff, 1978). A crowd pen for pigs must be designed with an abrupt entrance to the race. A round crowd pen with an abrupt entrance to the single-file race is being successfully used in several US pig abattoirs: two crowd gates continually revolve (Fig. 11.9). Another design is a single, offset step equal to the width of one pig. It prevents jamming at the entrance of a single-file race (Grandin, 1982, 1987). Jamming can be further prevented

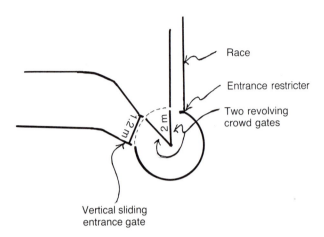

Race

Entrance restricter

Two revolving
crowd gates

Vertical sliding
entrance gate

Fig. 11.9. Round crowd pens for pigs with an abrupt
entrance to prevent jamming.

by installing an entrance restricter at single-file race entrances. The entrance
of the single-file race should provide only 1–2 cm on each side of each pig.
A double race should also have a single offset step to prevent jamming. In
Denmark an excellent double race system has been built with an offset step
on both sides of the double race (Barton-Gade, 1989). A power push gate
moves along the alley to urge the pigs into the races. Power push gates must
never be used to forcefully push animals, and they must be designed to prevent
animals from getting feet jammed under them.

For all species, solid sides are recommended on both the race and the
crowd pen which leads to the race (Brockway, 1977; Grandin, 1980a, b;
Grandin, 1982; Rider *et al.*, 1974; Vowles *et al.*, 1984). For operator safety,
mangates must be constructed so that people can escape charging cattle. The
crowd gate should also be solid to prevent animals from turning back. Wild
animals tend to be calmer in facilities with solid sides. In holding pens, solid
pen gates along the main drive alley facilitate animal movement (Grandin,
1980b).

When young pigs were given a choice of ramps, they preferred a ramp
with either solid or woven wire sides (Phillips *et al.*, 1987). Ramps with vertical
or horizontal barred sides were avoided. The overhead lighting used in the
indoor experiment may have made the wire mesh appear solid.

A crowd pen must never be built on a ramp. The animals will pile
up against the crowd gate. A small drainage slope will not affect
animal movement. To facilitate entry into the single-file race, construct two
to three body lengths of level single-file ramp before the animals reach the
ramp.

Unloading

In abattoirs more than one unloading ramp is usually required to facilitate prompt unloading. During warm weather, prompt unloading is essential because heat rapidly builds up in stationary vehicles. In some facilities, unloading pens (Fig. 11.1) will be required. These pens enable animals to be unloaded promptly prior to sorting, weighing or identification checking. After one or more procedures are performed, the animals move to a holding pen.

Loading dock height is going to vary depending on the types of vehicles used. If vehicle heights vary a few centimetres, it is recommended to construct non-adjustable ramps level with the lowest vehicles used: this will enable the cross-over bridge that is attached to the higher vehicles to be used more effectively. Facilities used for unloading only should be 2.5 m to 3 m wide to provide the animals with a clear exit into the alley (Fig. 11.1; Grandin, 1980d).

Ramps and slopes

Ideally an abattoir stockyard should be built at truck deck level to eliminate ramps for both unloading and movement to the stunner. Sheep move most easily on a level surface (Hitchcock and Hutson, 1979), and many animals are injured on excessively steep ramps. For fattened slaughter-weight pigs a 15° angle is recommended (VanPutten, 1981), but the maximum angle for non-adjustable livestock unloading ramps is 20° (Grandin, 1979). If possible, the ramp to the stunner should not exceed 10° for pigs, 15° for cattle, and 20° for sheep. A pig's heart rate increases as the angle of the ramp increases (VanPutten and Elshof, 1978), and the heart rate of a pig is faster when it is climbing a ramp compared to descending (Mayes and Jesse, 1980). Excessively steep ramps were avoided by pigs in a preference test; an angle of 20–24° was preferred compared to 28–32° (Fraser *et al.*, 1986; Phillips *et al.*, 1988). To reduce the possibility of falls, unloading ramps should have a flat dock at the top. This provides a level surface for animals to walk on when they first step off the truck (Agriculture Canada, 1984, Grandin, 1979; Stevens and Lyons, 1977), and this same principle also applies to ramps to the stunner. A level portion facilitates animal entry into the restrainer or stunning box.

Stairsteps are recommended on concrete ramps; they are easier to walk on after the ramp becomes worn or dirty. However, in new clean facilities, small pigs expressed no preference between stairsteps or closely spaced cleats (Phillips *et al.*, 1987). The movements in this experiment were voluntary.

Recommended dimensions for stairsteps on unloading ramps are a 30 cm minimum tread width and a 10 cm rise for cattle, and a 25 cm tread width and a 5 cm rise for slaughter-weight pigs (Grandin, 1980d, 1982; United States

Department of Agriculture, 1967). On a ramp to the stunner the tread width should be increased to 45 cm. The steps should be grooved to provide a non-slip surface. When cleats are used, space them 20 cm apart for large cattle and slaughter-weight pigs (Mayes, 1978). The 20 cm is measured from the beginning of one cleat to the beginning of the next cleat.

Design of restraint systems

Restraint devices to hold animals during stunning have improved since the days of large stunning boxes which hold more than one animal. One of the first major innovations was the V-conveyor restrainer for pigs which was patented in the USA by Regensburger in 1940. This device consists of two obliquely angled conveyors that together form a V-shape, and pigs ride with their legs protruding through the space at the bottom of the V (Fig. 11.10). In the late 1970s the Nijhuis company in Holland developed an automatic stunner incorporated into two V-restrainers: one restrainer runs faster than the other to index the pigs for the stunner. The V-restrainer is a humane system for most types of pigs and sheep (Grandin, 1980d); the pressure exerted against the sides of a pig will cause it to relax (Grandin *et al.*, 1989). However, the V-restrainer is not suitable for restraining extremely heavy muscled pigs with large overdeveloped hams; the V pinches the large hams and the slender forequarters are not supported.

For smaller pig abattoirs the squeeze box restrainer works well. It was patented in the USA by Hlavacek *et al.* in 1963 and consists of two padded panels which squeeze the pig. After stunning, the animal is ejected. In the 1980s a modified version of this device was developed for small European abattoirs. In the late 1960s the V-restrainer was enlarged for use on cattle by Oscar Schmidt of the Cincinnati Butcher's Supply Company and Don Williams of Armour and Company in the USA. A complete description and dimensions can be found in Grandin (1980d, 1983b).

The development of the V-restrainer for adult cattle was a major innovation because it replaced dangerous multiple cattle stunning boxes in high-speed slaughter plants. After stunning, the shackle chain was attached to one rear leg while the animal was still held in the restrainer.

V-restrainer systems are suitable for fat cattle, but there are problems with small calves and thin animals. Small calves tend to cross their legs and fall through (Giger *et al.*, 1977). Lambooy (1986) also reported that 200 kg veal calves had difficulty entering the restrainer.

Researchers at the University of Connecticut developed a laboratory prototype for a new type of restrainer system (Giger *et al.*, 1977; Westervelt *et al.*, 1976). Calves and sheep are supported under the belly and the brisket by two moving rails. This research demonstrated that animals restrained in

Fig. 11.10. V-restrainer conveyor for pigs.

this manner were under minimal stress. Sheep and calves rode quietly on the restrainer and seldom struggled. The space between the rails provides a space for the animal's brisket. The prototype was a major step forward in humane restrainer design, but many components still had to be developed to create a system which would operate under commercial conditions. Some of the items that needed to be developed were an entrance device which would reliably position the animal's legs on each side of the double rail, a rapid adjustment system and compatibility with existing abattoir equipment.

In 1986 the first double-rail restrainer was installed in a large commercial calf and sheep slaughter plant (Figs 11.11, 11.12 and 11.13). Dimensions and

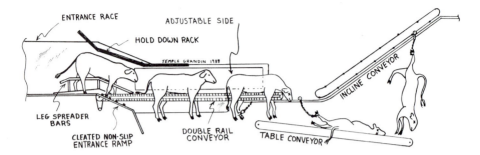

Fig. 11.11. Side view of the double-rail restrainer system.

details of the system can be found in Grandin, 1988. Development of the commercial system was accomplished by Grandin Livestock Handling Systems, Inc. and Clayton H. Landis Company.

In 1989 a V-restrainer conveyor was replaced by a double-rail restrainer in a large cattle slaughter plant by Grandin Livestock Handling Systems and Swilley Equipment, Logan, Iowa (Grandin, 1991)(Fig. 11.14). There are now eight large cattle systems operating in commercial slaughter plants. The double-rail restrainer has many advantages compared to the V-restrainer. Stunning is easier and more accurate because the operator can stand 28 cm closer to the animal. Cattle enter more easily because they can walk in with their legs in a natural position. Shackling is facilitated because the legs are spread apart and cattle ride more quietly in the double-rail restrainer.

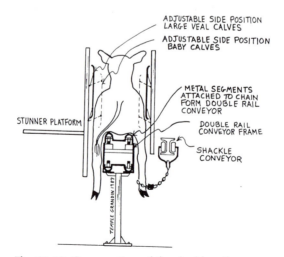

Fig. 11.12. Cross-section of the double-rail restrainer.

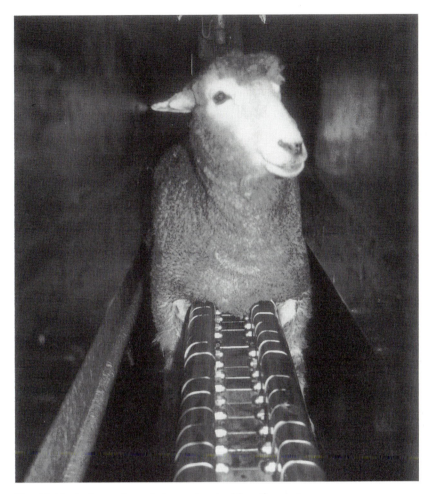

Fig. 11.13. A sheep sits quietly on the double-rail restrainer.

Proper design is essential for smooth humane operation. Incoming cattle must not be able to see light coming up from under the restrainer. It should have a false floor to provide incoming cattle with the appearance of a solid floor to walk on. The false floor is located about 20 cm below the animal's feet. The cattle restrainer also has a longer hold-down rack than the calf version. To keep cattle calm they must be fully restrained and settled down on the conveyor before they emerge from under the hold-down rack. Having a long enough hold-down rack is very important. If the hold-down is too short the cattle are more likely to become agitated. On both the calf and cattle versions of the double-rail restrainer there should be about 5 cm of clearance between an entering animal's back and the hold-down rack.

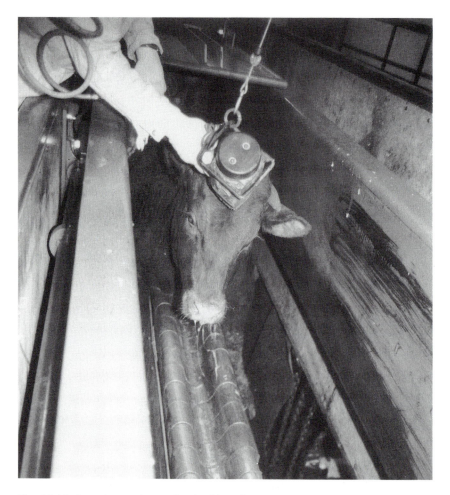

Fig. 11.14. Stunning cattle on the double-rail restrainer.

Conventional stunning boxes

The use of a device to hold the head during stunning is not required in either
a V-restrainer or a double-rail restrainer. In a conventional stunning box
stunning accuracy can be greatly improved by the use of a yoke to hold the
head. Yokes and automatic head restraints for cattle have been developed
in Australia, New Zealand and the UK. A common mistake is to build
stunning boxes too wide. A 76 cm wide stunning box will hold all cattle with
the exception of some of the largest bulls. Stunning boxes must have

non-slip floors; humane stunning and handling is impossible if an animal is skidding and slipping in the stunning box.

Ritual slaughter

Ritual slaughter is increasing in many countries due to increased Moslem demand for Halal meat. In some countries, such as the USA, it is legal to suspend live animals by one back leg for ritual slaughter. This cruel practice is also very dangerous. Replacement of shackling and hoisting by a humane restraint device will greatly reduce accidents (Grandin, 1988). In Europe and Australia the use of humane restraining devices is required.

The first restraining device for ritual slaughter was developed in Europe over 40 years ago. The Weinberg casting pen consists of a narrow stall which slowly inverts the animal until it is lying on its back. It is less stressful than shackling and hoisting, but it is much more stressful than more modern upright restraint devices (Dunn, 1990). Animals restrained in the Weinberg had much higher levels of vocalizations and cortisol (stress hormone) compared to cattle restrained in the upright ASPCA pen.

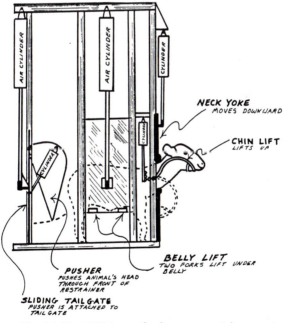

Fig. 11.15. ASPCA pen for humane upright restraint for ritual slaughter.

A major innovation in ritual restraint equipment was the ASPCA pen. It was developed in 1963 at Cross Packing Company in Philadelphia (Marshall *et al.*, 1963). It consists of a narrow stall with an opening in the front for the animal's head. A lift under the belly prevents the animal from collapsing after the throat cut (Fig. 11.15). Proper design and operation is essential. The belly lift should not lift the animal off the floor. Air pressure which operates the rear pusher gate should be reduced to prevent excessive pressure on the rear, and the head holder must have a stop or a bracket to prevent excessive bending of the neck.

Further, developments in ritual slaughter equipment are the use of a mechanical head holder on a V-restrainer (Grandin, 1980d) and ritual slaughter of calves on the double rail (Grandin, 1988). The research team at the University of Connecticut has also developed a small inexpensive restrainer to hold calves and sheep during ritual slaughter (Giger *et al.*, 1977). For larger calves a miniature ASPCA pen could be built.

Conclusions

The use of more humane equipment and handling practices will improve animal welfare. It will also have the added advantage of improving both meat quality and employee safety.

References

Agriculture Canada (1984) Recommended code of practice for care and handling of pigs. Publication 1771/E, Agriculture Canada, Ottawa.

Ames, D. R. (1974) Sound stress and meat animals. *Proceedings*, International Livestock Environment Symposium ASAE. SP-0174, 324.

Barton-Gade, P. (1985) Developments in the pre-slaughter handling of slaughter animals. *Proceedings*, European Meeting of Meat Research Workers. Albena, Bulgaria. Paper 1: 1, 1–6.

Barton-Gade, P. (1989) Pre-slaughter treatment and transportation in Denmark. *Proceedings*, International Congress of Meat Science and Technology, Copenhagen, Denmark.

Brockway, B. (1977) *Planning a Sheep Handling Unit*. Farm Buildings Centre, Kenilworth, Warwickshire, England.

Canadian Meat Council (1980) *Guide to PSE Pork*. Canadian Meat Council, Islington, Ontario.

Dunn, C. S. (1990) Stress reactions of cattle undergoing ritual slaughter using two methods of restraint. *Veterinary Record* 126, 522–5.

Eldridge, G. A. (1988) The influence of abattoir lairage conditions on the behavior and bruising of cattle. *Proceedings*, 34th International Congress of Meat Science and Technology, Brisbane, Australia.

Fraser, D., Phillips, P. A. and Thompson, R. K. (1986) A test of a free-access two level pen for fattening pigs. *Animal Production* 42, 269–74.

Giger, W., Prince, R. P., Westervelt, R. G. and Kinsman, D. M. (1977) Equipment for low stress animal slaughter. *Transactions of the American Society of Agricultural Engineers* 20, 571–8.

Grandin, T. (1979) Designing meat packing plant handling facilities for cattle and hogs. *Transactions of the American Society of Agricultural Engineers* 22, 912–17.

Grandin, T. (1980a) Observations of cattle behavior applied to design of cattle handing facilities. *Applied Animal Ethology* 6, 19–31.

Grandin, T. (1980b) Livestock behavior as related to handling facility design. *International Journal of the Study of Animal Problems* 1, 33–52.

Grandin, T. (1980c) Bruises and carcass damage. *International Journal of the Study of Animal Problems* 1, 121–37.

Grandin, T. (1980d) Designs and specifications for livestock handling equipment in slaughter plants. *International Journal of the Study of Animal Problems* 1, 178–200.

Grandin, T. (1982) Pig behavior studies applied to slaughter plant design. *Applied Animal Ethology* 9, 141–51.

Grandin, T. (1983a) Welfare requirements of handling facilities. In: Baxter, S. H., Baxter, M. R. and MacCormack, J. A. D. (eds), *Farm Animal Housing and Welfare*. Martinus Nijhoff, Boston/The Hague/Dordrecht/Lancaster.

Grandin, T. (1983b) System for handling cattle in large slaughter plants. American Society of Agricultural Engineers, Paper No. 83-406, ASAE, St Joseph, MI, USA.

Grandin, T. (1984) Race system for slaughter plants with 1.5 m radius curves. *Applied Animal Behavior Science* 13, 295–9.

Grandin, T. (1987) Animal handing. In: Price, E. O. (ed.), *Farm Animal Behavior*, vol. 3. Veterinary Clinics of North America, W. B. Saunders, Philadelphia, 323–38.

Grandin, T. (1988) Double rail restrainer conveyor for livestock handling. *Journal of Agricultural Engineering Research* 41, 327–38.

Grandin, T. (1989a) 'Effect of rearing environment and environmental enrichment on behaviour and neural development in young pigs.' Ph.D. dissertation, University of Illinois, Champaign/Urbana.

Grandin,T. (1989b) Behavioral principles of livestock handling. *The Professional Animal Scientist*, American Registry of Professional Animal Scientists, Champaign, Illinois, 5, 1–11.

Grandin, T. (1991) Double rail restrainer for handling beef cattle. American Society of Agricultural Engineers, Paper No. 91-5004, ASAE, St Joseph, MI, USA.

Grandin, T., Dodman, N. and Shuster, L. (1989) Effect of Naltrexone on relaxation induced by flank pressure in pigs. *Pharmacology, Biochemistry and Behavior* 33, 839–42.

Hitchcock, D. K. and Hutson, G. D. (1979) The movement of sheep on inclines. *Australian Journal of Experimental Agriculture and Animal Husbandry* 19, 176–82.

Hlavacek, R. J. (1963) Method for restraining animals, US Patent Number 3,115,670.

Hutson, G. D. (1981) Sheep movement on slotted floors. *Australian Journal of Experimental Agriculture and Animal Husbandry* 21, 474–9.

Hutson, G. D. (1985) Sheep and cattle handling facilities. In: Moore, B. L. and Chenoweth, P. J. (eds), *Grazing Animal Welfare*. Australian Veterinary Association, Queensland, 124–36.

Kenny, F. J. and Tarrant, P. V. (1987) The behavior of young Friesian bulls during social regrouping at an abattoir. Influence of an overhead electrified wire grid. *Applied Animal Behavior Science* 18, 233–46.

Kilgour, R. (1971) Animal handling in works; pertinent behavior studies. In: *Proceedings of the 13th Meat Industry Research Conference*, Hamilton, New Zealand, 9–12.

Kilgour, R. (1976) The behavior of farmed beef bulls. *New Zealand Journal of Agriculture* 13, 31–3.

Kilgour, R. (1978) The application of animal behavior and the humane care of farm animals. *Journal of Animal Science* 46, 1479–86.

Kilgour, R. (1983) Using operant test results for decisions on cattle welfare. *Proceedings*, Conference on the Human–Animal Bond, Minneapolis, Minnesota.

Kilgour, R. (1988) Behavior in the pre-slaughter and slaughter environments. *Proceedings*, International Congress of Meat Science and Technology, Part A, Brisbane, Australia, 130–8.

Kilgour, R. and deLangen, H. (1970) Stress in sheep from management practices. *Proceedings*, New Zealand Society of Animal Production 30, 65–76.

Lambooy, E. (1986) Automatic electrical stunning of veal calves in a V-type restrainer. *Proceedings*, 32nd European Meeting of Meat Research Workers, Ghent, Belgium, Paper 2:2, 77–80.

Lemman, W. B. and Patterson, G. H. (1964) Depth perception in sheep: Effects of interrupting the mother neonate bond. *Science* 145, 835–6.

Livestock Conservation Institute (1988) *Livestock Trucking Guide*, 6414 Copps Avenue, Madison, Wisconsin.

Lynch, J. J. and Alexander, G. (1973) *The Pastoral Industries of Australia*, University Press, Sydney, 371–400.

Marshall, B. L. (1977) Bruising in cattle presented for slaughter. *New Zealand Veterinary Journal* 25, 83–6.

Marshall, M., Milburg, E. E. and Shultz, E. W. (1963) Apparatus for holding cattle in position for humane slaughtering. US Patent 3,092,871.

Mayes, H. F. (1978) Design criteria for livestock loading chutes. ASAE paper 78-6014.

Mayes, H. F. and Jesse, G. W. (1980) Heart rate data of feeder pigs. ASAE paper 80-4023.

Meat and Livestock Commission (no date) Cattle handling. Livestock Buildings Consultancy, Meat and Livestock Commission, Queensway, Bletchley, Milton Keynes, England.

Midwest Plan Service (1980) *Structures and Environment Handbook*, 10th edn. Iowa State University, Ames, 319.

Phillips, P. A., Thompson, B. K. and Fraser, D. (1987) Ramp designs for young pigs. American Society of Agricultural Engineers, Paper No. 87-4511, St Joseph, MI, USA.

Phillips, P. A., Thompson, B. K. and Fraser, D. (1988) Preference tests of ramp designs for young pigs. *Canadian Journal of Animal Science* 68, 41–8.

Phillips, P. A., Thompson, B. K. and Fraser, D. (1989) The importance of cleat spacing in ramp design for young pigs. *Canadian Journal of Animal Science* 69, 483–6.

Puolanne, E. and Aalto, H. (1981) The incidence of dark cutting beef in young bulls in Finland. In: Hood, D. E. and Tarrant, P. V. (eds), *The Problem of Dark Cutting Beef*. Martinus Nijhoff, The Hague, 462–75.

Price, M. A. and Tennessen, T. (1981) Pre-slaughter management and dark cutting carcasses in young bulls. *Canadian Journal of Animal Science* 61, 205–8.

Prince, J. H. (1977) The eye and vision. In: Swenson, M. H. (ed.), *Dukes Physiology of Domestic Animals*. Cornell University Press, New York, 696–712.

Regensburger, R. W. (1940) Hog stunning pen. US Patent 2,185,949.

Rider, A., Butchbaker, A. F. and Harp, S. (1974) Beef working, sorting and loading facilities. American Society of Agricultural Engineers, Paper No. 74-4523, St Joseph, MI, USA.

Stevens, R. A. and Lyons, D. J. (1977) Livestock bruising project: Stockyard and crate design. National Materials Handling Bureau, Department of Productivity, Australia.

Stricklin, W. R., Graves, H. B. and Wilson, L. L. (1979) Some theoretical and observed relationships of fixed and portable spacing behavior in animals. *Applied Animal Ethology* 5, 201–14.

Tennessen, T. and Price, M. A. (1980) Mixing unacquainted bulls: The primary cause of dark cutting beef. *The 59th Annual Feeder's Day Report*. Agriculture and Forestry Bulletin, University of Alberta, Alberta, Canada, 34–5.

Tulloh, N. M. (1961) Behavior of cattle in yards. II: A study of temperament. *Animal Behavior* 9, 25–30.

United States Department of Agriculture (1967) Improving services and facilities at public stockyards. *Agriculture Handbook* 337, Packers and Stockyards Administration, United States Department of Agriculture, Washington, DC.

VanPutten, G. (1981) Handling slaughter pigs prior to loading and unloading the lorry. Paper presented at the Seminar on Transport, CEC, Brussels, 7–8 July.

VanPutten, G. and Elshoff, W. J. (1978) Observations on the effect of transport on the well being and lean quality of slaughter pigs. *Animal Regulation Studies* 1, 247–71.

Vowles, W. J., Eldridge, G. A. and Hollier, T. J. (1984) The behavior and movement of cattle through forcing yards. *Proceedings*, Australian Society for Animal Production 15, 766.

Warriss, P. D., Kestin, S. C. and Robinson, J. M. (1983) A note on the influence of rearing environment on meat quality in pigs. *Meat Science* 9, 271–9.

Westervelt, R. G., Kinsman, D. M., Prince, R. P. and Giger, W. (1976) Physiological stress measurement during slaughter of calves and lambs. *Journal of Animal Science* 42, 831–4.

Weyman, G. (1987) 'Unloading and loading facilities at livestock markets.' M.Sc. Thesis, Hatfield Polytechnic, UK.

Chapter 12

Biological Waste Treatment

W. Baader

Introduction

Agricultural wastes are in general the various organic residues which remain as vegetal biomass in harvesting plant products and in preparing these products for direct use and for sale, or which occur in animal husbandry as excrement more or less mixed with water and different solid matter (Table 12.1). If these organics cannot be put back immediately into the natural cycle, to be used for feeding, fertilizing or soil conditioning, and therefore have to be stored for later utilization, then measures have to be taken to avoid health risks and environmental pollution, caused by uncontrolled decay of organic substances. On the other hand, with suitable methods of treating and processing the waste properties can be improved, both with regard to handling and also to upgrade their value as fertilizer.

Table 12.1. Typical agricultural wastes.

Crop residues		Animal wastes	
Dry matter	Moist matter	Solid manure	Liquid manure
Straw	Leaves	Poultry excreta	Slurry:
Stalks	Green stalks	Animal excreta:	mixture of
Husks	Roots	mixed with	animal
Seed hulls	Grass	bedding	excreta and
Chaff	Peelings	matter (e.g.	water with
Bark	Press cake	straw, litter,	low con-
Shells	Pulp	sawdust) to	centration of
	Skins	stackable	solid
		state	additives
		Rumen content	(e.g. bedding
			matter)

Depending on the problems, risks and constraints connected with hand-ling, using or disposing of the wastes, different and occasionally combined aims for treatment and processing result:

1. Mineralization and stabilization:
   - prevention of uncontrolled biochemical decay of organic matter, connected with emissions of malodours and toxic gases and with the formation of mould fungi and other harmful substances.

2. Hygienization (sterilization):
   - deactivation of pathogenic organisms and parasites.

3. Adjustment of nutrients:
   - reduction of nutrient concentration to avoid water pollution due to overfertilizing during land-spreading;
   - recovery and/or enrichment of nutrients to provide fertilizers for con-trolled application;
   - purification of liquid being discharged into rivers;
   - conversion of compounds to achieve a higher nutritive value.

4. Improvement of handling:
   - reduction of viscosity,
   - transformation of liquid to a homogeneous suspension, easily pumped and suitable for controlled and exact land-spreading;
   - transformation of solids to a loose material suitable for stacking, piling and spreading.

5. Reduction of mass and volume.

In achieving these aims, biological processes are of great importance (Fig. 12.1). In all biological processes, organic matter is decomposed, disintegrated or converted to lower polymeric compounds by means of micro-organisms. The basic conditions for these microbiological metabolisms are the avail-ability of nutrients, the observance of adequate ranges for pH and tem-perature, and the presence of enough moisture to ensure the microbial functions.

In general, all agricultural wastes are suitable for being degraded by micro-organisms, because their essential compounds are carbohydrates, fats, proteins, cellulose and hemicellulose. Lignin, however, is resistant to micro-bial decay. Therefore biomass containing cellulose encrusted by lignin has to be pre-treated to break open this binding. By means of biological treatment, agricultural wastes can be converted to different products which are suitable for re-use in agriculture (Fig. 12.2).

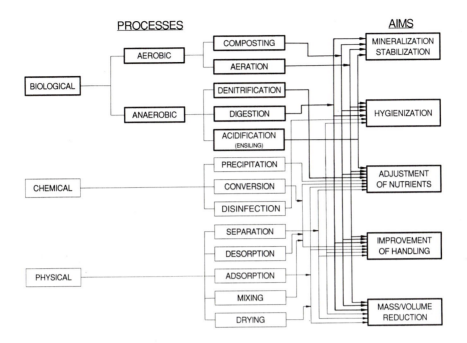

Fig. 12.1. Processes and aims of treating agricultural wastes (with particular consideration of biological treatment).

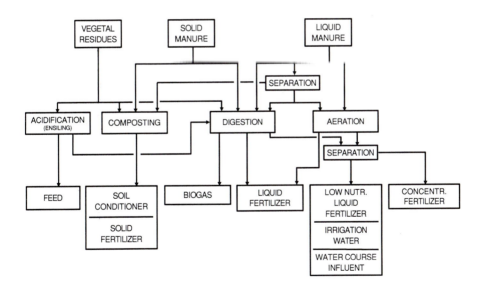

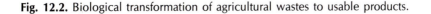

Fig. 12.2. Biological transformation of agricultural wastes to usable products.

Principles of biological treatment

Aerobic treatment of solids (composting)

Composting is an exothermic aerobic process in which organic compounds of solid matter are degraded by organisms like bacteria, actinomycetes and fungi, to carbon dioxide, nitrates, ammonia and water. Additionally, biomass and ulmous substances are formed. Certain environmental factors have to be provided to ensure an accurate process performance. The most important factors are the moisture content and the porosity of the solid substrate (Schuchardt, 1987a).

The moisture content should be the maximum in the range in which water is held, and a minimum of not less than about 35% w.b. Suitable measures for adjusting the moisture content to its optimum value are the addition of water, slurry or organic sludges (such as liquid manure) if the substrate is not moist enough, or − if the substrate is too wet − the addition of water-adsorbing organic matter (such as chopped or sliced straw, bark, sawdust and wood chips), or the separation of surplus liquid. The porosity of the substrate is responsible for supplying the organisms with oxygen by procuring air, and for discharging the metabolism products carbon dioxide, ammonia and water vapour. The porosity needed can be reached if structured solid matter is used. If the original substrate is too dense, it has to be mixed with suitable loosening materials (Schuchardt, 1987a, b).

During the composting process, heat is produced by the microbial oxidation of carbon (34−42kJ/gC). In larger heaps in which the heat can accumulate as a result of insulation, the temperature can rise to more than 75°C. The highest rate of degradation occurs at temperatures between 50°C and 60°C. Sterilization can be achieved at temperatures of at least 60°C lasting for a minimum of one day (Fachem *et al.*, 1983). But with regard to release of ammonia the temperature should be less than about 40°C.

In cases where the continuing decomposition of the material causes the porosity to decrease, resulting in a reduction of the gas exchange followed by limitation of the process, the substrate heap must be loosened, remixed or turned frequently by mechanical aids; or the oxygen supply can be boosted by forced aeration.

Aerobic treatment of liquids (aeration)

Organic matter which is dissolved or suspended in liquids can be degraded in a biological process comparable to composting, provided that the micro-organisms are supplied with enough oxygen. Therefore it is necessary to distribute air in the liquid in such a way that very small bubbles are produced; thus, a large surface ensures an efficient transfer of oxygen into the liquid phase. Different systems of aeration are in practical use. Most designs combine both air intake and liquid mixing (Cumby, 1987).

In agriculture, the main purpose of aeration is to reduce the odour of liquid manure in order to prevent impairment of the environment. This effect is achieved by the biological degradation of odorous compounds of the manure. A further effect of this oxidation process is the production of heat. In insulated systems, and if the concentration of degradable organic matter is high enough, the temperature can rise to 60°C; thus pathogens are deactivated. This advantage of sterilization is accompanied by a release of ammonia, which means a reduction in the value of the final liquid fertilizer (Thaer and Grabbe, 1976; Thaer, 1978).

If the temperature does not exceed about 35°C, the ammonia is converted by bacteria to nitrates. If the aim of treatment is a major reduction in the content of nitrate in the final liquid, which may then be utilized for irrigation or be released into a water course, a further step of biological denitrification is necessary, in which the nitrate is reduced to nitrogen by bacteria under anaerobic conditions, and the nitrogen is released into the atmosphere.

Anaerobic digestion

The process of anaerobic digestion is characterized by the microbial degradation of complex polymeric organic compounds under anaerobic conditions in neutral or slightly alkaline conditions, and in temperatures ranging from 25°C to 55°C. Carbon and hydrogen elements from the original matter are combined with the metabolism products carbon dioxide and methane in a mixture released as biogas. The total process is effected in several sequential steps. In the first step, the polymers are degraded by hydrolytic and acid-forming bacteria to compounds of low molecular weight, which are further transformed by acetic-acid-forming and hydrogen-forming bacteria. In the final step, specialized bacteria produce methane from hydrogen and acetic acid, accompanied by carbon dioxide (Sahm, 1981). Normally, the different steps of the total process are accomplished in one reaction space. However, if complex wastes like vegetal residues are used, a two-phase process may be of advantage (Verstraete *et al.*, 1981). In this mode, the hydrolysis/acid-forming step and the methane-forming step are physically separated. Thus by pH-controlled feeding of the second step, limitation of the activity of the methane-forming bacteria resulting from too high a concentration of acids is prevented.

High yields of methane are achievable from wastes rich in carbohydrates (such as sugar and starch), fats or proteins. From this follow the differences of the related data from typical agricultural wastes (Table 12.2).

Table 12.2. Methane production from typical agricultural wastes.

Type of waste	Methane yield (l/kg vol. solids)	Methane concentration in biogas (%)
Cow manure	180–250	60–70
Pig manure	210–300	58–60
Poultry manure	350–400	58–65
Rumen content	160–300	60–65
Green plants	250–450	55–62
Straw	150–180	60–62
Potato pulp	270–300	58–60
Fruit press cake	300–450	60–65

Advanced biological treatment processes

Composting of organic sludges

Organic sludges are characterized by a largely homogeneous structure of fine particle solids, a moisture content resulting in a high viscosity, and few or no air-containing spaces. Typical agricultural wastes of this kind are animal manures without bedding matter added. For composting it is necessary to transform the sludge into a porous structure and to adjust the moisture content to the optimum range. In conventional composting systems, the sludge is mixed with additional structuring solid matter, and the mixture has to be loosened frequently during the period of composting. Procuring, preparing (for example, by size reduction) and handling these solid additives, as well as the increase of the total mass which has to be treated, are factors which considerably increase the expense.

In an advanced composting system, (Fig. 12.3), developed at the FAL-Institute of Technology, Braunschweig, Germany (Baader *et al.*, 1979), the moisture content of the original sludge is adjusted to about 50% either by adding dry and crushed end product, or, if manure from caged laying hens is processed, by in-building aeration. With this moisture content a plastic and cohesive material is achieved which can be pressed easily and with a low energy requirement into aggregates of, preferably, about 50mm in diameter and 50–100mm in length (Fig. 12.4) – strong enough to be heaped up to 3.5m in height without being destroyed by pressure. When these aggregates are placed into a bin or a tower silo, equipped with a perforated bottom and an open top, the aerobic decomposing process commences spontaneously, generating heat which induces a convectional vertical air-flow through the total heap due to the homogeneous porous structure formed by the aggregates. From this process the system is self-supplied with oxygen, thus neither

POULTRY
MANURE   COMPOST

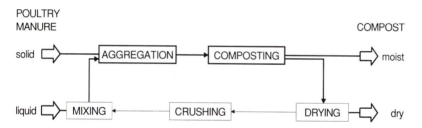

Fig. 12.3. Composting of aggregated poultry manure – basic flow diagram.

Fig. 12.4. Aggregates formed from poultry manure of 50% dry matter content.

forced and energy consuming aeration nor loosening and/or turning of the substrate are necessary. The aggregates are formed with an axial hole (Fig. 12.5) in order to prevent internal anaerobic uncontrolled processes, which emit bad odours. Because of the high temperatures, in the range 50–70°C, generated by the thermophilic process within the total mass of substrate during at least 5–7 days residence, it is guaranteed that all pathogenic organisms are deactivated. The compact modular construction of the reactor makes it possible to collect the ammonia-concentrated emission for further processing, to protect the environment and to produce a nitrogenous fertilizer.

Fig. 12.5. Press tool for forming aggregates with an axial hole from sludge enriched in solids.

Aerobic thermophilic removal of nitrogen from liquid manure

In principle, liquid manure is considered to be a valuable organic fertilizer, but if applied in quantities in excess of the actual nutrient demand of the plants, the risk of environmental impairment by groundwater pollution with organic matter and nitrates can occur, depending on the local soil conditions. In cases where the amount of liquid manure produced in an individual animal enterprise cannot be disposed to the fields in relation to the actual nutrient demand of the plants, the manure has to be transported over longer distances to be spread over larger areas – thus, however, incurring relatively high costs due to the large water content. This problem can be minimized or solved by reduction of the concentration of nutrients, first of all nitrogen as ammonium, so that more liquid per area can be spread locally, while the nutrients removed, and subsequently concentrated, are more suitable for transport.

As mentioned above, ammonia is released from ammonium-containing manure, if an aerobic thermophilic process is encouraged. This effect can be used for producing a low-nitrogen liquid fertilizer, combined with recovering the nitrogen as a concentrate, in which the ammonia is fixed by acid or water, that can be used like a mineral fertilizer.

It is known from experimental work (Thaer, 1975) and also from operating farm-scale plant, in which the aerobic thermophilic treatment process has been used for stabilization, sterilization and reduction of malodour, that a

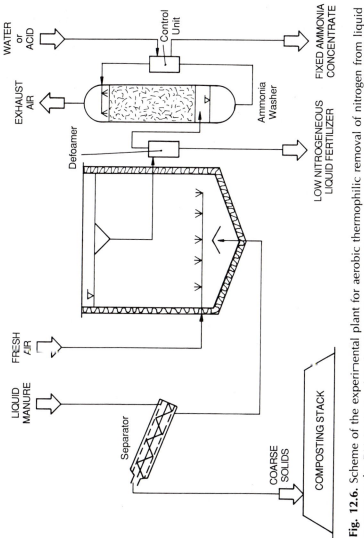

Fig. 12.6. Scheme of the experimental plant for aerobic thermophilic removal of nitrogen from liquid manure at the FAL-Institute of Technology, Braunschweig.

reduction of nitrogen content in liquid manure of up to 50% is attainable. It can be expected that by optimizing the process conditions, the efficiency of the ammonia desorption will increase, thus resulting in a lower nitrogen content in the final liquid fertilizer. In order to determine the optimum conditions and the resulting maximum nitrogen removal rate, and also to evaluate the economics of this way of treating liquid manure, a project was started in 1990 at the Institute of Technology, FAL Braunschweig.

The main component of the experimental equipment used in this project is a cylindrical insulated and covered tank-reactor ($1m^3$ volume), fitted with aeration units at the bottom. With a screen press the coarse particles are separated from the liquid manure and subsequently composted. The liquid substrate is fed continuously to the reactor and, after a hydraulic retention time of up to approximately ten days, the treated liquid is discharged, together with gas, water vapour and foam. After the 'defoamer', the gas and the water vapour pass through the ammonia washer, in which the ammonia is fixed either by water or by acid as a concentrate. The schematic and simplified performance of the experimental plant is shown in Fig. 12.6.

High-rate anaerobic digestion of liquid manure

The efficiency of an anaerobic treatment process depends not only on the provision of the essential conditions for performing the process, but also and in particular on the degradability of the substrate. Liquid manure is a mixture of water, dissolved organic and inorganic compounds, and suspended solids of different shape and size. Whilst most of the dissolved matter is easily biologically degradable, most of the suspended solids need a comparatively longer time for being degraded because they consist predominantly of polymeric complex organic matter. If a high yield of gas is expected from the complete mixture, the retention time for the substrate in the reactor is determined by the degradation time for the poorest degradable matter. In completely mixed throughflow reactors – the conventional type in common use (Demuynck *et al.*, 1984) – the retention time for the suspended biomass (solids and bacteria) is equal to the hydraulic retention time.

The efficiency of an anaerobic digester can be increased if the suspended biomass is retained in the reactor for a longer retention time, combined with accumulation of bacteria; whilst for the liquid the (hydraulic) retention time is adjusted to achieve fast degradation of the dissolved matter. Of advantage is the accumulation of bacteria in the reactor by sedimentation as granules, floccules, or adjoined to solids (upflow anaerobic sludge blanket (UASB) reactor), or by deposition and adhesion at the surface of stationary carrier structures (fixed bed reactor), or of mobile carrier particles (expanded bed or fluidized bed reactor); see Fig. 12.7. The stationary carriers used are wide porous, randomly packed beds of inert particles or well-structured plastic

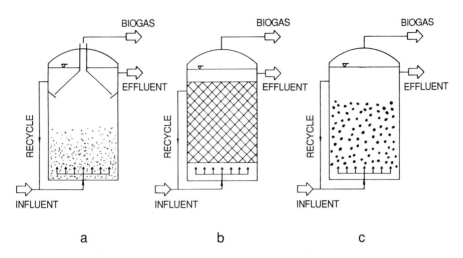

Fig. 12.7. High-rate anaerobic digestion systems with biomass accumulation:
(a) upflow anaerobic sludge blanket reactor (UASB); (b) fixed bed reactor;
(c) fluidized bed reactor.

beds of inert particles or well-structured plastic material, both of large surface area per volume, which fill out the reactor volume, normally at 80–90% (Weiland, 1987).

In expanded bed and in fluidized bed reactors, floccular or granular inert particles of a specific gravity less than that of the substrate are placed in the reactor, where they are mobilized by upflowing liquid or ascending gas bubbles. The movement of the particles is controlled by the vertical velocity of flow, obtained by liquid or gas recirculation. In the expanded state, all particles keep their position within the loosened bed, whereas in the fluidized state the single particle is moved over longer distances (Weiland and Büttgenbach, 1988). It is evident that high-rate anaerobic digesters equipped with biomass-retaining carriers are very sensitive to suspended, and in particular to coarse solids in the substrate. Therefore in order to prevent clogging, uncontrolled solid accumulation and the formation of scum layers from floated matter, solids which may cause such risks have to be separated before the substrate is fed to the digester.

In pilot-scale investigations at the author's laboratory a fixed bed reactor and a UASB reactor (each of 1.3 m^3 net volume and 0.6 m diameter) were compared with a completely mixed reactor. As substrate, liquid pig manure was used, from which solids of more than 2 mm size have been separated by screening. With the completely mixed digester (conventional system), the optimum yield of gas could be obtained with a hydraulic retention time of 20 days. With both high-rate digesters, however, the same yield was reached with a retention time of only five days.

Biomethanation of solid wastes

The conventional way of anaerobic digestion of solid wastes is to mix the material after crushing as an additive to liquid manure or similar organic slurries used as the main substrate. In this case, the biogas plant is part of a liquid handling and treating system. However, if the vegetal matter is the predominant substrate, every additional liquid quantity may cause costs and management needs to rise, both to provide the liquid as well as to handle and to dispose of the effluent. In order to eliminate these disadvantages, instead of adding fresh liquid to the solid material before feeding it into the digester, either the effluent or liquid separated from the effluent may be recycled.

Digesting of vegetal matter in the fluid state brings with it the problem of flotation. Therefore measures must be provided to prevent formation of floating layers in the upper part of the digester, which may cause blockage in the material flow and decrease efficiency.

These problems can be solved with the horizontal type digester (Fig. 12.8(a)), in which the fluid is smoothly stirred transversely by means of slowly rotating paddles which push floating matter into the fluid. In the vertical tank type digester, the floatable matter must also be controlled by mechanical means. Two designs have proved to be suited for handling fluid substrates with high concentrations of vegetal matter. In the system developed at the Agricultural Research Centre, Invermay, New Zealand, the sludge is moved up through a vertical tube by a propeller, and is spread on to the floated matter (Fig. 12.8(b)). The digester is fed intermittently with a mixture of chopped vegetal matter after liquid of the same volume has been taken from the lower part of the digester (Stewart, 1983). The system developed at the FAL-Institute of Technology, Braunschweig, uses a completely filled reactor with a conical shape at both bottom and top (Fig. 12.8(c)). In the upper section of the reactor a slowly rotating screw agitator is installed in a central tube which acts downwards, transferring floating material into deeper zones of the reactor. The substrate is fed into the lower section of the reactor and is mixed simultaneously with both recycled liquid from the reactor effluent and sludge taken from the reactor bottom. The anaerobically treated waste and gas leave the digester at the uppermost point and are fed to a gas separating and scum destroying unit. The reactor effluent is de-watered by a screw-type filter press and some of the liquid phase is recycled to the reactor for diluting the feed. The system can be used with a variety of wastes without scum formation. It necessitates a fluid phase with a total solids content up to about 10% for fibrous solids, and more for free-flowing solids of small particle size (Baader, 1985).

The Dry Anaerobic Composting (DRANCO) system (Fig. 12.8(d)), developed at the University of Ghent, Belgium, is operated at an extremely high total solids content of more than 20% with downflow of the semi-solid

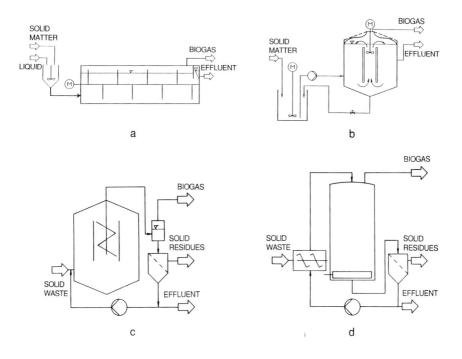

Fig. 12.8. One-stage systems for the anaerobic digestion of solid wastes:
(a) horizontal plug-flow, radially stirred reactor; (b) completely stirred reactor
(ARC system, Invermay); (c) partially stirred reactor (FAL system, Braunschweig);
(d) unstirred downflow reactor (DRANCO system, University of Ghent).

phase and no use of agitation. The wastes are fed to the top of the reactor.
The anaerobically treated solids are withdrawn from the bottom of the reactor
using a special scraper, and are subsequently de-watered by a chamber press.
The liquid phase is mostly recycled. The solids are dried or composted (De
Baere and Verstraete, 1984; Six and De Baere, 1988).

In two-phase digestion systems, solid wastes are first microbially pre-
treated in a separate hydrolysis reactor to achieve an extensive liquefaction
and acidification of the solids. Biomethanation of the liquid phase taken from
the hydrolysis reactor is then accomplished in a second reactor, which can
be operated with biomass accumulation, if the non-liquefied solids are removed from the feed. Figure 12.9 shows two different two-phase digester
systems, one developed at the Institute for Storage and Processing of
Agricultural Products (IBVL) at Wageningen in the Netherlands (Hofenk,
1986), and the other developed at the FAL-Institute of Technology (Albin
et al., 1990).

In the IBVL process, the solid waste matter is liquefied and acidified
batchwise in the hydrolysis reactor, then the liquid is percolated through the

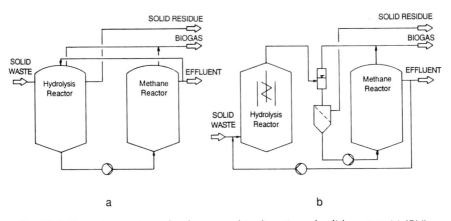

Fig. 12.9. Two-stage systems for the anaerobic digestion of solid wastes: (a) IBVL system, Wageningen; (b) FAL system, Braunschweig.

packed bed of solid waste to leach the soluble acidic organic compounds. The leachate is then fed to a UASB reactor for converting the dissolved organic compounds mainly to methane. From the effluent of the methanogenic stage one part is recirculated to the hydrolysis stage and used for leaching.

The FAL process avoids the disadvantages of a batchwise operation of the hydrolysis stage by using a completely filled, partially stirred hydrolysis reactor, which is continuously fed. This reactor is of similar design to that used in one-phase operation for biomethanation of solid substrates (see Fig. 12.8(c)). The solid waste is mixed with effluent from the methane stage and fed into the lower section of the hydrolysis reactor. The liquefied and acidified slurry leaves the reactor at the uppermost point and is de-watered in a subsequent screen press. The liquid phase is fed to the methane stage, which can be operated with different methods of biomass accumulation. The solid residues can be composted, and most of the effluent from the methane reactor is recycled to the hydrolysis stage. It is obvious that the FAL system can be operated with solids of different physical properties, both structured and unstructured wastes.

Biological processes integrated in waste treatment systems: some examples

Processing of poultry manure

The system of composting manure with high dry matter content (achievable, for example, by indoor aeration drying) using the thermophilic aerobic treatment of aggregated matter, as described above, is realized in the manure treatment plant at the Gutshof-Ei poultry farm at Schwege, Germany, where the manure from 120,000 caged laying hens is processed to compost for sale (Fig.12.10). The manure leaves the building with about 50–60% of dry matter and is subsequently transformed to aggregates (by means of a screw-type extruder press) which are delivered to the composting bins. The plant is equipped with eight batchwise operated bins of 28m^3 volume each, which are filled successively. After eight days of biological treatment during which the temperature increases up to 70°C, caused by the microbial exothermal oxidation and conversion process, the content of each individual bin is aerated for about two days by means of a mobile low-pressure fan (Fig. 12.11). The product, with about 80% d.m., is free of pathogen risks. It can be used as fertilizer and is sold either in bulk state, or in bags after further processing by crushing, pelletizing and drying, suitable for application in horticulture.

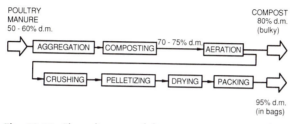

Fig. 12.10. Flow diagram of the manure treatment plant at the Gutshof-Ei poultry farm.

Centralized composting of solid agricultural wastes

If solid residues are not recycled at the farm, either because there is a surplus of that material, which may cause environment problems, or because there is an external demand for organic fertilizer and soil conditioner, then composting the wastes for sale is of growing interest. In that case the product has to meet the restrictions of authorities for environmental protection and health. Also, the acceptance by the consumer depends on the quality of the product, which has to be declared and guaranteed, often on the basis of an official certification. To meet these obligations would overwhelm an individual farm and requirements can be better met by a central processing plant, to which the various solids from the surrounding farms are transported. Therefore a successfully working solid waste treatment system with integrated composting process is described next.

In the Backhus plant, located near Bösel, Germany, various solid wastes are processed to a compost of defined quality. The wastes – predominantly manure from turkeys, broilers, laying hens, pigs and cows, and solids separated from liquid dairy manure, as well as spoiled silage and hay, straw, rice husks, slaughterhouse manure and rumen waste, are collected by the

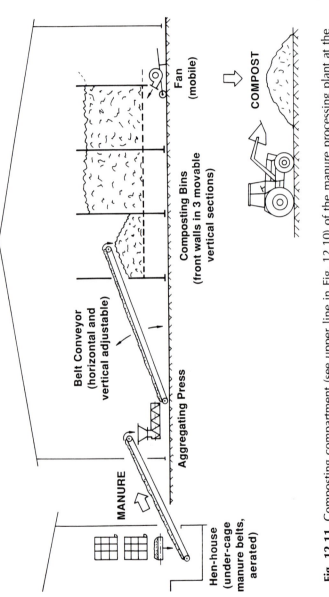

Fig. 12.11. Composting compartment (see upper line in Fig. 12.10) of the manure processing plant at the Gutshof-Ei poultry farm

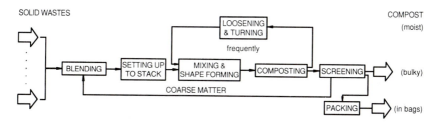

Fig. 12.12. Flow diagram of the Backhus composting plant.

unit's own vehicles or delivered by contractors. The processing stages are shown by the flow diagram in Fig. 12.12. To reach the optimum ratios of the different substances with regard to the subsequent composting process, the various wastes are blended and set in heaps of 2–3 m in height by means of a wheel-loader. After about two weeks of storage, during which a first aerobic exothermal degradation process occurs, the matter is transported with the loader into the composting hall and set up in stacks, in which intensive and controlled aerobic treatment is performed over 4–6 weeks.

Using a self-propelled machine (Fig. 12.13), the matter is picked up, intensively mixed and set up in an accurate trapezoid shape of 1.8 m height and 2.5/0.5 m width. This action is repeated three times a week to accomplish an optimum composting process by loosening and turning the matter. The raw compost is subsequently screened, and the coarse fraction is recycled to the blending area. The fine fraction, a crumbly, approximately 50% moist and earthy-smelling product of defined – and declared – composition and free of pathogens, is sold either in bulk, or after having passed a weighing and packing station, in bags of 20–80 l volume for use in individual and commercial horticulture.

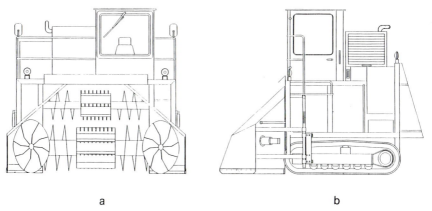

a   b

Fig. 12.13. Self-propelled machine for stack forming, and for substrate turning and mixing (the Backhus system 'Kompostmat'): (a) front view; (b) side view.

Centralized treatment of liquid manure

In rural areas, where due to a high concentration of animal production there is the risk of groundwater pollution by nitrates penetrating the deeper zones of the soil, and where the problem cannot be solved by such measures as:

● increasing the storage capacity to enable manure disposal at optimum time and quantity for fertilizing;
● distributing the manure over longer distances to enlarge the area for correct disposal;
● reducing the nutrient concentration in the liquid by separating solids which then can be used outside the farm; or
● reducing the number of animals;

then the liquid manure has to be treated with the aim of considerable reduction of the polluting substances. In the most critical case it may be necessary to remove organic substances and minerals to such an extent that the final liquid can be discharged into the river. This aim can only be achieved with an integrated treatment system, which consists of several components of sequential steps for treating the matter in physical, biological and chemical processes.

In Germany the first plant incorporating this concept, built by the Sulzer Company at Haverbeck near Damme, Germany, was put into operation in 1990. The plant, which is on a pilot scale, processes liquid piggery waste of about $10,000\,m^3$ a year to a purified effluent which, according to the restrictions of the local environment protection authority, may be discharged to the nearby river (Fig. 12.14). The liquid manure is delivered to the plant from the surrounding family farms. The substances removed from the manure are transformed to products worthy of being sold as fertilizer.

The configuration of the treatment plant follows the flow diagram in Fig. 12.15. In the first step, the liquid manure is subjected to an alkaline treatment (Chemolyse), by which ammonia is removed and complex compounds are disintegrated to enhance the efficiency of the following separation. The solid fraction is dried to an organic fertilizer, and the liquid fraction is fed to the anaerobic digester of $600\,m^3$ volume where, along with conversion of organic matter to biogas, the nitrogen fixed in organic compounds is converted to ammonium. This biological treatment is followed by a step in which phosphate is precipitated, so that it can be separated from the liquid as calcium phosphate. The nitrogen desorbed as ammonia is removed by stripping and subsequently, together with the ammonia released in the first step, transformed by absorption with acid to ammonium sulphate or to ammonium phosphate. The reduction of the residual organic load in the effluent in order to reach the restricted value, is accomplished in the final aerobic treatment step. The biogas is converted to heat and electricity from which a considerable part of the energy required for the process can be obtained.

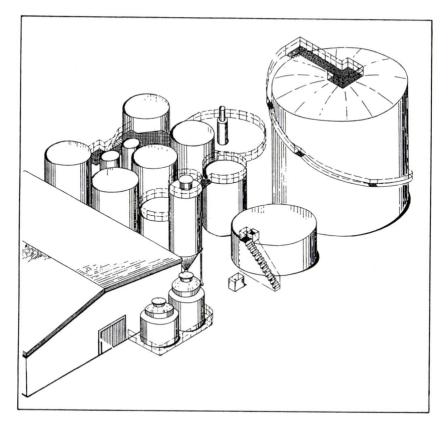

Fig. 12.14. Outline drawing of the Sulzer plant for piggery waste treatment.

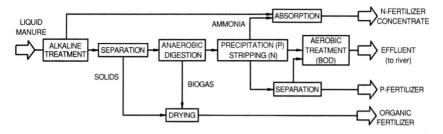

Fig. 12.15. Flow diagram of the Sulzer plant for piggery waste treatment.

Biological treatment of slaughterhouse wastes

At slaughterhouses, different kinds of wastes are encountered. All of these consist predominantly of biodegradable substances which decay rapidly and, if they are not handled and treated under controlled conditions, cause severe problems with regard to the environment and hygiene risks. Wastes in this category include solid and moist matter like rumen contents, manure and substances separated from the waste water by screening and flotation, and waste water highly concentrated with both dissolved and suspended organic matter. In order to get rid of them, in most cases the solids have been deposited at municipal refuse landfills, and the waste water discharged into the municipal sewers. However, because the capacity of both the landfills and the sewage treatment plants is limited and the costs for these ways of waste management are increasing, systems must be developed by which both environmental pollution and costs can be reduced.

One activity with that aim is an R&D project at the FAL Institute of Technology where, on a pilot-scale, an integrated system for treatment of solid and liquid slaughterhouse wastes is under successful investigation (Fig. 12.16).

As is shown in the flow diagram (Fig. 12.17), the solid matter, predominantly rumen content (waste of about 15% d.m. which is delivered from the local municipal slaughterhouse), is in the first stage treated in an anaerobic digester ($28\,m^3$ volume) of the completely mixed and totally filled type described earlier and shown in Fig. 12.8(c). The solid residues are separated from the effluent by means of a screw-type screen press and subsequently composted, either in stacks or in bins. The residual liquid is partially recycled to the digester for diluting the substrate to about 8% d.m., in order to ensure an accurate free flow of matter inside the digester. The excess amount of

Fig. 12.16. Pilot plant for slaughterhouse waste treatment at the FAL Institute of Technology, Braunschweig.

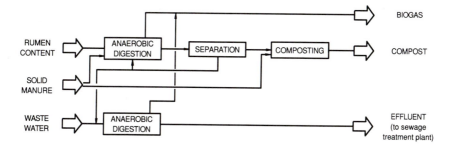

Fig. 12.17. Flow diagram of the FAL pilot plant for slaughterhouse waste treatment.

liquid is added to the waste water. Also, solid manure from the slaughterhouse lairage can be fed to the digester, or it can be composted directly.

The waste water, also derived from the local slaughterhouse, is anaerobically treated in an upflow digester (2.8 m³ volume) of the fixed bed type (Fig. 12.7(b)). In this stage the organic load is reduced by up to about 90% so the effluent can be delivered to a sewage treatment plant where only the nitrogen concentration has to be reduced, at a lower cost than for treating the total original waste water.

The advantages of the system are:

● with regard to the solid waste:
  - reduction of odour;
  - reduction of mass and volume;
  - production of hygienic compost for use as soil conditioner and fertilizer;
  - production of biogas for use in the slaughterhouse;

● and with regard to the waste water:
  - reduction of odour;
  - reduction of organic load, which results in reduction of costs for final treatment;
  - production of biogas.

References

Albin, A., Ahlgrimm, H. J. and Weiland, P. (1990) Biomethanation of solid and semi-solid residues from harvesting and processing of renewable feedstocks. In: Grassi, G., Gosse, G. and dos Santos, G. (eds), *Proceedings of 5th EC Conference: Biomass for Energy and Industry*. Vol. 2. Elsevier Applied Science Publishers, London and New York, 132–8.

Baader, W. (1985) Performance of a completely filled vertical through-flow anaerobic digester. International Symposium *Alternative Sources of Energy for Agriculture.* Book Series No. 28, Food and Fertilizer Technology Center, Taipei, 51–65.

Baader, W., Schuchardt, F. and Sonnenberg, H. (1979) Transformation of liquid manure into a solid. *Agricultural Wastes* 1, 167–90.

Cumby, T. R. (1987) A review of slurry aeration. *Journal of Agricultural Engineering Research* 36, 141–206.

De Baere, L. and Verstraete, W. (1984) Anaerobic fermentation of semi-solid and solid substrates. In: Ferrero G. L., Ferranti, M. P. and Naveau, H. (eds), *Anaerobic Digestion and Carbohydrate Hydrolysis of Waste.* Elsevier Applied Science Publishers, London and New York, 195–208.

Demuynck, M., Nyns, E. J. and Palz, W. (1984) Biogas Plants in Europe. *Solar Energy R & D in the European Community. Series E: Energy from Biomass*, Vol. 6. D. Reidel, Dortrecht, Holland.

Fachem, R. G., Bradley, D. J., Garelick, H. and Mara, D. D. (eds) (1983) *Sanitation and Disease-Health Aspects of Excreta and Wastewater Management.* John Wiley, Chichester, UK.

Hofenk, G. (1986) Anaerobic digestion of organic wastes in one and two stage processes. In: Baader, W. (ed.), *Biomass Conversion for Energy.* CNRE bulletin No. 10a, Food and Agriculture Organization of the United Nations, Rome, 17–21.

Sahm, H. (1981) Biologie der Methanbildung (Biology of methanogenesis). *Chemie-Ingenieuer-Technik* 53, 854–63.

Schuchardt, F. (1987a) Composting of Liquid Manure and Straw. In: Welte, E. and Szaboles, I. (eds), 4th International CIEC Symposium *Agricultural Waste Management and Environmental Protection.* Federal Research Center of Agriculture (FAL), Braunschweig, 271–81.

Schuchardt, F. (1987b) Zur Bedeutung des Luftporenvolumens für die Kompostierung organischer Schlämme (Significance of the free space for the composting of organic sludges). *Grundlagen der Landtechnik* 37, 108–15.

Six, W. and De Baere, L. (1988) Dry anaerobic composting of various organic wastes. In: Tilche, A. and Rozzi, A. (eds), 5th International Symposium *Anaerobic Digestion.* Monduzzi Editore, Bologna, 793–7.

Stewart, D. J. (1983) Methane from crop-grown biomass. In: Wise, D. L. (ed.), *Fuel Gas Systems.* CRC Press, Boca Raton, FL, 85–109.

Thaer, R. (1975) Behandlung von Rinderflüssigmist (Treatment of liquid dairy manure). *Berichte über Landwirtschaft*, Special Issue 192, 836–81.

Thaer, R. (1978) Probleme der aeroben Behandlung von Flüssigmist in flüssiger Phase (Problems of aerobic treatment of liquid manure). *Grundlagen der Landtechnik* 28, 36–47.

Thaer, R. and Grabbe, K. (1976) Flüssigmistfermentation mit Selbsterwärmung (Liquid manure fermentation with spontaneous heating). *Grundlagen der Landtechnik* 26, 215–21.

Verstraete, W., De Baere, L. and Rozzi, A. (1981) Phase separation in anaerobic digestion. *La Tribune de CEBEDEAU* 34, 367–75.

Weiland, P. (1987) Development of anaerobic filters for treatment of high strength agro-industrial wastewaters. *Bioprocess Engineering* 2, 39–47.

Weiland, P. and Büttgenbach, L. (1988) Study of different fluidized bed reactors using porous plastic materials as supports. In: Tilche, A. and Rozzi, A. (eds), 5th International Symposium *Anaerobic Digestion*. Monduzzi Editore, Bologna, 345–9.

Index